Student Study Guide

for use with

Microbiology

A Human Perspective

Fourth Edition

Eugene W. Nester
Denise G. Anderson
C. Evans Roberts, Jr.
All of the University of Washington

Nancy N. Pearsall
Martha T. Nester

Prepared by
Redding I. Corbett III
and
Michael Lema
both of Midlands Technical College

 **Higher Education**

Boston   Burr Ridge, IL   Dubuque, IA   Madison, WI   New York   San Francisco   St. Louis
Bangkok   Bogotá   Caracas   Kuala Lumpur   Lisbon   London   Madrid   Mexico City
Milan   Montreal   New Delhi   Santiago   Seoul   Singapore   Sydney   Taipei   Toronto

The McGraw-Hill Companies

Student Study Guide for use with
MICROBIOLOGY: A HUMAN PERSPECTIVE, FOURTH EDITION
NESTER/ANDERSON/ROBERTS/PEARSALL/NESTER

Published by McGraw-Hill Higher Education, an imprint of The McGraw-Hill Companies, Inc.,
1221 Avenue of the Americas, New York, NY 10020. Copyright © The McGraw-Hill Companies,
Inc., 2004, 2001, 1998, 1995. All rights reserved.

3 4 5 6 7 8 9 0 QPD QPD 0 9 8 7 6 5 4

ISBN 0-07-247505-6

www.mhhe.com

Table of Contents
Microbiology: A Human Perspective, 4e
Nester, Anderson, Roberts, Pearsall, Nester

I. LIFE AND DEATH OF MICROORGANISMS

II. THE MICROBIAL WORLD

III. MICROORGANISMS AND HUMANS

IV. INFECTIOUS DISEASES

V. APPLIED MICROBIOLOGY

Preface

As you stand on the threshold of your study of the microbiological world, you are about to begin an exciting odyssey. You may be anxious, expectant, and we hope, excited. You have a fascinating world to explore. Learning is just that, exploration and discovery, not rote memorization of unrelated facts. You will discover a new world – a world too small to be seen with the unaided eye and you will learn how microorganisms play a role in your life either for good or for ill. As you journey through this fascinating world, let this study guide be your map.

This study guide is not intended to replace your textbook or lectures, nor is it a shortcut to learning. It is designed to help you organize your study, to help you to identify and learn the important points and to help you identify weaknesses in your understanding so that you might address them before you have a test.

Each chapter of the study guide begins with a short overview of the textbook chapter. This is intended to focus you on the topic presented. It is followed by a section titled Key Concepts, which restates the major concepts from the chapter in a concise direct manner. These sections along with your textbook chapter should be read prior to the presentation of the material in class.

The Learning Objectives specifically direct you to material that you will need to know. There are two ways to use these objectives. First, you may use them as a guide for learning, and second you may use them as a checklist to make sure that you have not overlooked something important.

The Summary Outline is different from the one found in your textbook. This abbreviated outline is intended to direct your attention toward terms, concepts, microorganisms and diseases that are important both in learning the text material and in doing the study guide activities.

Every discipline has its own language and science is no exception. Learning the language is basic to learning the concepts. You should define, preferably in writing, each term in the Terms You Should Know section. The List of Microorganisms and the List of Diseases are included where appropriate. These lists should help you focus on important organisms that you may also encounter in other chapters and on diseases caused by these microorganisms.

The last three sections are the active learning sections. They are designed to help you use your knowledge by (1) evaluating what you learned from the chapters, (2) creating associations, and (3) understanding concepts. The Self Test section offers you the opportunity to see if you comprehend the material. The Learning Activities are not just for evaluation, but for helping you to understand concepts and create associations. The Thought Questions are designed to give you an opportunity to apply your knowledge and to think about the applications and implications of this material. While answer guidelines are provided to aid you in answering the Thought Questions, your answers may differ or be more expansive. That is OK. It is good to be creative and thoughtful. As you journey through this odyssey, discover, think, and enjoy your adventure.

Solving Critical Thinking Problems

Included at the end of each chapter of your textbooks are critical thinking exercises that provide practice for applying the concepts and information included in the chapter. These exercises extend well beyond the correct recall of information or "looking up the answer." They require you to utilize critical thinking skills such as interpreting data and experimental results, predicting outcomes when conditions are changed, proposing and evaluating experimental designs, and establishing sound arguments and lines of reasoning. In other words, the emphasis is on skill development and application.
In each exercise, you should develop a logical argument that is based on information in the chapter and sound reasoning, not on opinion. The strength of your argument and reasoning will depend on appropriate concept application and structuring your argument so that it clearly leads to your conclusion and/or interpretation.

A few general guidelines for completing the exercises will be helpful:

1) Know what the question is asking. Does the question ask for a prediction? for an interpretation? for an experimental analysis?

2) What information is needed to solve the exercise? Now is the time for you to go back through the chapter, if necessary, to review concepts and information. Be sure to apply appropriate information to the problem.

3) What new information is provided in the exercise? How does this exercise differ from examples included in the chapter? Decide how this new information fits with and extends information in the chapter.

4) Draw a diagram or outline of the exercise. This will help organize your thoughts and insure understanding of the relationships in the exercise. It will also indicate the structure of your logic leading to your solution.

For most students, these exercises are difficult, especially at first. But as with any skill, ability and skill will improve with practice. Do not be discouraged with early difficulties; consistent effort and practice will lead to significant improvement.

The text website provides answers to the critical thinking questions included in the Microchecks in your textbook. The more you practice solving these exercises, the better your skills will become and you will become proficient in solving similar types of exercises in any discipline.

www.mhhe.com/nester4

Chapter 1 Humans and the Microbial World

Overview

This chapter is an introduction to microbiology with a discussion of some of the major events that have been important in the development of the science. The origin and discovery of microorganisms is explored and the concept of two cell types, prokaryotic and eukaryotic, is introduced. The impact of microorganisms on humans is explored as well as the roles of these organisms in the biological world. The three domains are introduced and various kinds of microorganisms included in them are described. Obligate intracellular parasites, which are not included in the domains, are introduced as well as binomial nomenclature. The chapter concludes with a discussion of microbial size.

Learning Objectives

After studying the material in this chapter, you should be able to:

1. Define microbiology and describe its scope.
2. Define microorganism
3. Describe the environmental conditions under which the first microbes lived.
4. Explain why microorganisms have been successful.
5. Describe the diversity of microorganisms.
6. Briefly describe the contributions of the following people to the field of microbiology.
    *   Antony van Leeuwenhoek
    *   Francesco Redi
    *   Louis Pasteur
    *   John Tyndal
    *   Ferdinand Cohn
    *   Robert Koch
    *   Edward Jenner
    *   Mathias Schleiden and Theodor Schwann
    *   Ignas Semmelweiss
    *   Joseph Lister
    *   Dimitri Ivanowski
    *   Paul Ehrlich
    *   Frederick Griffith
    *   Alexander Fleming
    *   Joshua Lederberg and Edward Tatum
    *   James Watson, Francis Crick, Rosalind Franklin and Maurice Wilkins
    *   Luc Montagnier and Robert Gallo
7. Describe briefly the theory of spontaneous generation.
8. List the types of organisms encountered in the study of microbiology.
9. List ten important roles of microorganisms.
10. Distinguish between emerging diseases and re-emerging diseases and list some examples of each.
11. Explain why microorganisms are useful model organisms for study.

12. List the three domains and the major characteristics of each.
13. List and describe the kinds of microorganisms in each of the domains.
14. Demonstrate how to correctly write the scientific name of an organism.
15. Distinguish among virus, viroid and prion.
16. Describe the size range of microorganisms.

Key Concepts

1. Antony van Leeuwenhoek first observed microorganisms about 300 years ago.
2. Pasteur and Tyndall refuted the theory of spontaneous generation about 140 years ago.
3. Microorganisms are essential to all life on earth and impact the life of humans in both beneficial and harmful ways.
4. Microorganisms have been used for food production for thousands of years using essentially the same techniques as used today.
5. Microorganisms are now being used to degrade toxic pollutants and produce a variety of compounds more cheaply than can be done in the laboratory.
6. Through genetic engineering the capabilities of microorganisms have been greatly expanded.
7. Genomics is the study of the DNA sequences of organisms.
8. While great progress has been made in preventing and treating infectious diseases, new ones continue to arise around the world.
9. Microorganisms provide useful models for the study of biochemical principles and genetics.
10. Based on cell structure and chemical composition all organisms can be divided into three large groups: the Bacteria, the Archaea and the Eucarya.
11. The Bacteria and the Archaea are prokaryotic organisms identical in appearance with a simple cell structure but differ in chemical composition.
12. The Eucarya, which include algae, fungi, protozoa and multicellular parasites, have a complex cell structure and are termed eukaryotes.
13. All organisms are classified according to the Binomial System of Nomenclature.
14. Members of the microbial world that are acellular are the obligate intracellular parasites, viruses, viroids and prions.
15. Members of the microbial world vary greatly in size with obligate intracellular parasites usually being the smallest and eukaryotes the largest.

Summary Outline

1.1 The origin of microorganisms
    A. Theory of Spontaneous Generation
        1 Pasteur's experiments
        2 Tyndall and Cohn experiments
    B. The first microorganisms—probably grew in the absence of air and at very high temperatures
1.2 Microbiology: A human perspective
    A. **Vital activities** of microorganisms
        1 Necessary for the survival of all other organisms
        2 Bacteria fix nitrogen; microorganisms replenish the oxygen on earth
        3 Microorganisms degrade organic waste materials
    B. **Economic applications** of microbiology
        1 Production of bread, wine, beer and cheeses
        2 Bacteria degrade dangerous toxic pollutants
        3 Bacteria synthesize a variety of different products

C. **Genetic engineering**
    1   Genes from one organism are introduced into related or unrelated organisms resulting in new properties
    2   Expands the capabilities of microorganisms enormously
    3   Microorganisms produce medically important products including vaccines
    4   Genes can be transferred into plants by microorganisms

D. **Genomics**
    1   The science that deals with the DNA sequences of organisms
    2   Genomics will enable scientists to better understand the relationships between organisms and with their environments.

E. **Medical microbiology**
    1   Microorganisms cause diseases such as smallpox, bubonic plague and influenza
    2   Emerging diseases are arising in developed countries
    3   Other diseases that were declining have begun to reemerge
    4   Chronic diseases such as ulcers and heart disease may be caused by bacteria
    5   Bacteria use the body as an ecological niche

F. Microorganisms as **subjects for study**
    1   Excellent model organisms to study
    2   Grow rapidly and follow the same genetic, metabolic and biochemical principles as higher organisms

1.3 The Microbial World

A. Two major cell types
    1   The simple **prokaryotic**
    2   The complex **eukaryotic**

B. **Three domains**—based on the chemical composition and cell structures
    1   **The Bacteria**—single-celled prokaryotes with peptidoglycan in their cell wall
    2   **The Archaea**—single-celled prokaryotes; do not have peptidoglycan in their cell wall; grow in extreme environments
    3   **The Eucarya**—have eukaryotic cell structure: single cells or multicellular
    4   Microbial members of the Eucarya are
    5   **Algae**—single-celled or multicellular; can use sunlight as a source of energy
    6   **Fungi**—single-celled yeasts or multicellular molds and mushrooms; use organic compounds as food
    7   **Protozoa**—single-celled organisms
        a) Motile by a variety of means
        b) Use organic compounds as food

C. Nomenclature
    1   **Binomial system**
    2   *Genus* and a *species* name written in Italics

1.4 **Viruses, viroids and prions**

A. Non-living members of the microbial
B. Not composed of cells
C. **Obligate intracellular parasites**
D. **Prions** consist only of protein without any nucleic acid

1.5 **Size** in the microbial world—**varies greatly**

Terms You Should Know

Algae
Bioremediation
Biotechnology
Domains
Emerging diseases
Endospores
Eukaryotes

Fungi
Genetic engineering
Genomics
Genus
Host
Microorganisms
Pathogen

Prions
Prokaryotes
Protozoa
Species
Spontaneous generation
Viroids
Viruses

Learning Activities

1. How would you describe a microorganism to someone who had never heard of one?

2. Microorganisms have definite roles in the biological world. Make a chart listing their various roles and classify those roles as essential (life can't exist on earth without them), beneficial (life is better with them) or harmful to humans (life would be better without them).

Essential	Beneficial	Harmful

3. Compare prokaryotic and eukaryotic cells by completing the following table.

Characteristics	Prokaryotic Cell	Eukaryotic Cell
Cell size		
Nuclear membrane		
Cell wall		
Organelles such as mitochondria		

4. Match the following scientists with their major contributions.

C	Leeuwenhoek	A.	Proposed that all organisms are composed of cells.
	Snow	B.	Demonstrated that puerperal or childbed fever is a contagious disease transmitted by physicians to their patients during childbirth
	Schleiden and Schwann	C.	Invented the microscope
	Tyndall	D.	Developed evidence against spontaneous generation
	Griffith	E.	Introduced antiseptic surgery
	Watson, Crick, Franklin and Wilkins	F.	Verified Pasteur's evidence against spontaneous generation
	Pasteur	G.	Used salvarsan to treat syphilis
	Koch	H.	Discovered genetic transformation
	Semmelweiss	I.	Discovered penicillin
	Lister	K.	Developed the first vaccine
	Ehrlich	L.	Demonstrated the epidemic spread of cholera through water
	Jenner	M.	Identified the causative agent of tuberculosis
	Fleming	N.	Determined the structure of DNA

5. Describe the kinds of organisms in each domain.

Bacteria	Archaea	Eucarya

6.   Compare Bacteria, Archaea and Eucarya by completing the following table.

Characteristics	Bacteria	Archaea	Eucarya
Cell size			
Nuclear membrane			
Cell wall composition			
Mitochondria			
Chloroplasts			
Cytoskeleton			
Habitat (environment of the organism)			

7.   Match each organism with the domain in which it belongs.

C	1. Protozoa	A. Bacteria
	2. *Escherichia coli (bacteria)*	B. Archaea
	3. Yeasts	C. Eucarya
	4. Algae	D. Not in a domain
	5. Pine Tree	
	6. Bacteria	
	7. Molds	
	8. Viruses	
	9. Organisms that live in extreme environments	
	10. Mushrooms	

8. Compare the following types of eukaryotic microorganisms by completing the table.

	Algae	Fungi	Protozoa
Cell organization			
Source of energy			
Size (microscopic or macroscopic			

9. Compare viruses, viroids and prions by completing the following table.

	Viruses	Viroids	Prions
Obligate intracellular parasite			
Contains DNA only			
Contains RNA only			
Contains neither DNA or RNA			
Protein coat present			

Self Test

1. Leeuwenhoek's discovery was of great significance because

    a. he was a well-known scientist of his day.
    b. he developed a superior lens.
    c. he was the first person to use a microscope.
    d. he carefully recorded his observations and repeated his results.
    e. he was the first person to make observations using magnifying lenses.

2. Infectious diseases have essentially been eliminated by antibiotics and vaccination and no longer pose a public health problem.

    a. true
    b. false

3. Which of the following diseases were once thought to be controlled, but are now reappearing in developed countries?

    1. measles
    2. whooping cough
    3. AIDS
    4. tuberculosis
    5. mumps

    a. 1 2, 3, 4
    b. 2, 3, 4, 5
    c. 3
    d. 1, 2, 4, 5
    e. 1, 2, 3, 4, 5

4. The discovery of the microbial world by Leeuwenhoek created a controversy that lasted almost 300 years. This controversy centered on

    a. the cause of disease.
    b. spontaneous generation.
    c. the opposition of the church.
    d. the binomial system of nomenclature.

5. Tyndall and Cohn working separately discovered a dormant bacterial form that is heat-resistant. It is known as the

    a. vegetative cell.
    b. prion.
    c. endospore.
    d. exospore.
    e. cell wall.

6. What is the process in which bacteria are used to destroy dangerous chemical pollutants?

    a. biotechnology
    b. bioremediation
    c. genetic engineering
    d. transformation
    e. None of the above

7. Which of the following are domains?

    1. Bacteria
    2. Archaea
    3. Eucarya
    4. Prokarya

    a. 1, 2, 3
    b. 2 and 3
    c. only 3
    d. 2 and 4
    e. 1, 2, 3 and 4

8. Viruses cannot be classified as prokaryotes or eukaryotes because they

    a. are not living.
    b. are chemicals.
    c. were discovered after prokaryotes and eukaryotes.
    d. are not cellular.
    e. do not contain any genetic information.

9. Obligate intracellular parasites which consist of a piece of nucleic acid surrounded by a protein coat are

    a. viruses.
    b. viroids.
    c. prions.
    d. Both a and b are correct.
    e. All of the above are correct.

10. Which of the following is the correct way for a scientific name to appear in print?

    a. *Bacillus cereus*
    b. *Bacillus Cereus*
    c. *bacillus cereus*
    d. Bacillus cereus
    e. bacillus Cereus

Thought Questions

1. It is said that life could not exist on Earth without microorganisms. Why is this true?

2. It is said that we are in a new age of infectious disease. Why is there a resurgence of infectious disease?

3.   Data indicates that the rates of infectious disease are increasing. Explain why this is the case.

Answers to Self Test Questions

1-d, 2-b, 3-d, 4-b, 5-c, 6-c, 7-d, 8-a, 9-a, 10-a

Chapter 2 The Molecules of Life

Overview

All material things are based on chemicals. Organisms are no exception. To understand the function as well as the structure of microorganisms, it is necessary to understand their chemistry. This chapter presents the basic concepts of chemistry as it relates to the biological sciences, most especially microbiology. The chapter includes an introduction to atoms, bonding and the chemical components of a cell, including macromolecules. The structure and function of proteins, carbohydrates, lipids and nucleic acids are presented.

Learning Objectives

After studying the material in this chapter, you should be able to:

1. Describe the structure of an atom.
2. Calculate the number of protons, neutrons, and electrons from the atomic weight and atomic number of a given element.
3. Give the four most important elements in living organisms.
4. Explain why bonds are formed.
5. List and describe the types of chemical bonds. Give examples of each.
6. Explain why the bonding properties of water are important.
7. Explain the importance of pH.
8. Describe what occurs in the following chemical reactions.
   - Dehydration synthesis
   - Hydrolysis
9. List the four biologically important classes of organic molecules and their subclasses.
10. Describe the kinds of compounds included in each of the four biologically classes of organic molecules.
11. Differentiate among
    - Primary structure of a protein
    - Secondary structure of a protein
    - Tertiary structure of a protein
    - Quaternary structure of a protein
12. Identify the following
    - Denatured protein
    - Substituted or conjugated protein
13. Differentiate among
    - Monosaccharides
    - Disaccharides
    - Polysaccharides
14. List the components of a nucleotide.
15. List the nitrogenous bases in
    - DNA
    - RNA
16. Differentiate between saturated and unsaturated fatty acids.

Key Concepts

1. The four most important elements in living organisms are carbon, hydrogen, oxygen and nitrogen.
2. The atom is the basic unit of matter.
3. Atoms are composed of protons, electrons and neutrons.
4. Electrons are negatively charged particles within the energy levels or orbital of an atom.
5. Protons are positively charged particles within the nucleus of an atom.
6. Neutrons are uncharged particles within an atom the nucleus of an atom.
7. The atomic number is the number of protons in the nucleus of an atom that equals the number of electrons in the orbitals.
8. The atomic weight or mass is the number of protons and neutrons in the nucleus of an atom.
9. Molecules are formed by the creation of bonds between atoms.
10. Valence is the number of electrons that an atom can give-up, receive, or share.
11. Bonds are the forces that holds atoms together to form molecules.
12. Types of bonds include covalent, which are strong, and ionic and hydrogen, which are generally weak.
13. The polar bonds of water molecules are responsible for the many properties of water required for life on earth.
14. pH is a measure of the acidity or alkalinity of a substance.
15. Dehydration synthesis is the joining together of two or more molecules by the removal of water.
16. Hydrolysis is the splitting apart of two or more molecules by the addition of water.
17. Macromolecules consist of many repeating subunits, each subunit being similar or identical to the other subunits.
18. The side chains of amino acids are responsible for their properties.
19. Proteins are molecules that are composed of amino acids bonded together.
20. Carbohydrates are sugar molecules consisting of carbon, hydrogen and oxygen bonded in a ratio of 2 hydrogens for each carbon and oxygen and polymers of these sugars.
21. Carbohydrates perform a variety of functions in cells, including serving as a source of energy and forming part of the cell's structures.
22. Carbohydrates with the same composition of elements can have very different properties because of different arrangements of the atoms in the molecules.
23. Nucleic acids are polymers of nucleotide subunits that in turn consist of a pentose sugar, either ribose or deoxyribose, a phosphate group and a nitrogenous base, either adenine, cytosine, guanine, thymine (DNA only) or uracil (RNA only).
24. DNA carries the genetic code in the sequence of purine and pyrimidine bases in its double helical structure.
25. RNA transfers the genetic code information from DNA to the ribosome to make protein.
26. Lipids are heterogeneous hydrocarbon molecules containing little oxygen.
27. Phospholipids with one polar end and the other non-polar form a major part of cell membranes.

Summary Outline

2.1 Atoms and elements
    A. **Atoms** are composed of **electrons, protons** and **neutrons**
    B. An **element** is a pure substance.
2.2 Chemical **bonds** and the formation of molecules
    A. The outer orbital of electrons of an atom must be filled for maximum stability
    B. **Valence** is the number of electrons that an atom may gain, give up, share when bonds are formed

C. Bonds form between atoms to fill their outer orbitals
    1. **Covalent bonds**—strong bonds formed by atoms sharing electrons
        a) **Nonpolar**—equal attraction for electrons
        b) **Polar**—unequal attraction for electrons
    2. **Ionic bonds**—electrons leave the orbital of one atom and enter the orbital of another atom
    3. **Hydrogen bonds**—weak bonds that result from the attraction of a positively charged hydrogen atom in a polar molecule to a negatively charged atom in another polar molecule

2.3. Chemical components of the cell
  A. **Water**
    1. Most important molecule in the cell comprising 70% of all living organisms by weight
    2. Hydrogen bonding plays a very important role in the properties of water
  B. **pH** is the degree of acidity of a solution, measured on a scale of 0-14
  C. Small molecules in the cell
    1. Variety of small organic and inorganic molecules
    2. Carbon occurs in all organic molecules
    3. **Adenosine triphosphate (ATP)** is modified RNA nucleotide that is used to store energy within a cell and provides energy when the terminal bond is broken to form **adenosine diphosphate (ADP)** and inorganic phosphate
  D. Macromolecules and their component parts
    1. Composed of subunits with similar properties
    2. Synthesis of macromolecules occurs by **dehydration synthesis**
    3. Degradation occurs by **hydrolysis**

2.4 **Proteins**—most versatile of the macromolecules
  A. Activities of proteins include:
    1. Catalyzing reactions
    2. Being a component of cell structures
    3. Moving cells
    4. Taking nutrients into the cell
    5. Turning genes on and off
    6. Being a part of cell membranes
  B. **Amino acids**
    1. Proteins are composed of **20 major amino acids**
    2. All amino acids are carbon compounds containing both a carboxyl group and an amino group
    3. The side chain of the amino acid confers unique properties on the amino acid
  C. **Peptide bonds**—amino acids are joined through peptide bonds, joining amino with carboxyl groups and splitting out water
  D. Protein structure
    1. The **primary structure**—amino acid sequence
    2. The **secondary structure**—coiling and folding into helices and sheets
    3. The **tertiary structure**—the three-dimensional shape of the protein
    4. The **quaternary structure** interaction of several polypeptide chains
    5. Denaturation of proteins—intramolecular bonds within the protein are broken and the protein changes shape and no longer functions
  E. Substituted proteins—contain other molecules such as sugars and lipids, bonded to the side chains of amino acids in the protein.

2.5   **Carbohydrates**—heterogeneous group of compounds
    A.   Perform a variety of functions in the cell
        1.   Have carbon, hydrogen, and oxygen atoms in a ratio of approximately 1:2:1
    B.   **Monosaccharides**—classified by the number of carbon atoms they contain, most commonly 5 or 6
    C.   **Disaccharides**—two monosaccharides joined by dehydration synthesis
    D.   **Polysaccharides**—consist of monosaccharide subunits

2.6.   **Nucleic acids**—Macromolecules whose subunits are nucleotides
    A.   **DNA**—Carries genetic information of the cell in its sequence of nucleotides
        1.   Double-stranded helical molecule
        2.   Composed of covalently bonded sugar, phosphate group and a purine or pyrimidine base
        3.   The two strands are complementary and are held together by hydrogen bonds between the bases
    B.   **RNA**—involved in decoding the genetic information contained in DNA
        1.   Single-stranded molecule and contains uracil in place of thymine

2.7   **Lipids**—heterogeneous group of molecules that are only slightly soluble in water
    A.   **Simple lipids**—simple lipids contain carbon, hydrogen and oxygen and may be liquid or solid at room temperature
        1.   Fats consist of glycerol bound to fatty acids
        2.   Lipids may be saturated (contains no double bonds between carbon atoms) or unsaturated (contains one or more double bonds)
        3.   Some simple lipids consist of a four-membered ring
    B.   **Compound lipids**—contain elements other than carbon, hydrogen and oxygen; example—phospholipids

Terms You Should Know

Amino acid	Hydrolysis	Primary structure of a protein
Atom	Ionic bond	Protein
Atomic number	Lipid	Proton
Atomic weight	Macromolecule	Orbital
ATP	Monosaccharide	Quaternary structure of a protein
Buffer	Neutron	Ribonucleic acid (RNA)
Carbohydrate	Nonpolar covalent bond	Secondary structure of a protein
Dehydration synthesis	Nucleotide	Stereoisomer
Denature	Peptide bond	Structural isomer
Deoxyribonucleic acid (DNA)	pH	Tertiary structure of a protein
Electron	Phospholipid	Triglyceride
Element	Polar covalent bond	Valence
Hydrogen bond	Polysaccharide	

Learning Activities

1. Complete the following table about the basic structure of an atom.

Particle	Location	Charge	Weight or Mass
Proton			
Neutron			
Electron			

2. What does the atomic number represent? _____

3. What does the atomic weight represent? _____

4. Calculate the number of protons and neutrons in the nucleus of an atom and the number of electrons in the orbitals or energy levels from the atomic weight and atomic number of the elements given and complete the table below.

Element	Atomic Number	Atomic Weight	Number of Protons	Number of Neutrons	Number of Electrons
Hydrogen	1	1			
Carbon	6	12.			
Nitrogen	7	14			
Oxygen	8	15.9994			
Phosphorus	15	31			
Sulfur	16	31			

5. Match the type of bond with the appropriate statement.

	Bonds within a water molecule	A. Ionic bond
	Weak bonds that hold the two strands within a DNA molecule together	B. Nonpolar covalent bond
	Bonds within a CH_4 molecule	C. Polar covalent bond
	Bonds within NaCl molecule	D. Hydrogen bond
	Bonds between two water molecules	

6.   Complete the following table. Place a check mark next to the solution that is closest to neutral.

PH	Acidic or basic?
1. pH = 8.124	
2. pH = 4.518	
3. pH = 6.815	
4. pH = 9.128	
5. pH = 1.0386	
6. pH = 7.498	

7.   Indicate how many times more acidic or basic are the following compounds than pure water (pH=7.0).

1. Stomach acid – pH = 2.0 *	
2. Limewater – pH = 14.0	
3. Tomato juice – pH = 4.0	
4. Urine – pH = 6.0 *	
5. Household ammonia – pH = 11.0	
6. Wine – pH = 4.0 *	

* Actual values may vary

8.   Match the biologically important classes of organic molecules with the appropriate statement: (Choices may be used once, more than once or not at all.)

Class of Organic Molecule
A. Carbohydrates
B. Lipids
C. Proteins
D. Nucleic Acids
E. All of the above

	1. Composed of carbon, hydrogen and oxygen in the same ratio as water
	2. Composed of amino acids
	3. Contain carbon, hydrogen, oxygen, nitrogen and phosphorus
	4. Most abundant component of a cell membrane
	5. Composes genes—the hereditary material
	6. Contains carbon, hydrogen and a small amount of oxygen
	7. Includes starch, cellulose and glycogen
	8. Contains peptide bonds
	9. Formed by dehydration synthesis

9. Describe the following types of protein structure.

Primary structure	
Secondary structure	
Tertiary structure	
Quaternary structure	

10. Explain what happens when a protein is denatured.

11. List the specific components of DNA and RNA.

	DNA	RNA
Phosphate group		
Sugar		
Bases		

Self Test

1. The sodium ion, Na^+, has a single positive charge because it has

    a. more neutrons than electrons
    b. more protons than electrons
    c. more electrons than protons
    d. more electrons than neutrons

2. The number of protons in the nucleus of an atom is its

    a. atomic number.
    b. number of electron orbitals (energy levels).
    c. atomic weight.
    d. atomic mass.
    e. orbitals.

3. Atoms are electrically neutral. Which of the following possess an electrical charge?

    1. proton
    2. electron
    4. ion
    5. neutrons
    6. atoms

    a. 1, 3, 5
    b. 3, 4, 5
    c. 1, 2, 3
    d. only 3
    e. 1, 2, 3, 4, 5

4. Weak bonds that are responsible for holding the two strands of a DNA molecule together are

    a. ionic bonds.
    b. hydrogen bonds.
    c. disulfide bonds.
    d. nitrogen.
    e. covalent bonds.

5. Biological molecules made by covalently bonding amino acids together are called

    a. proteins.
    b. lipids.
    c. nucleic acids.
    d. disaccharides.
    e. polysaccharides.

6. Macromolecules are formed by joining their subunits through

    a. hydrolysis.
    b. dehydration synthesis.
    c. hydrogen bonding
    d. hydration synthesis.
    e. dehydrolysis.

7. The sequence of amino acids in a protein determines its

    a. primary structure.
    b. secondary structure.
    c. tertiary structure.
    d. quarternary structure.

8. Lactose and sucrose are examples of

    a. DNA.
    b. monosaccharides.
    c. proteins.
    d. polysaccharides.
    e. disaccharides.

9. Amino acids consist of which of the following?

    a. an amino (NH_2) group
    b. a side chain ("R" group)
    c. a carboxyl (COOH) group
    d. Both a and b.
    e. Both a, b and c.

10. The carbon to hydrogen to oxygen ratio of carbohydrates is

    a. very large.
    b. 1:2:1.
    c. 1:2:2.
    d. 2:1:4.
    e. variable.

11. DNA differs from RNA in that DNA has

    1. two strands
    2. deoxyribose
    1. one strand
    2. ribose
    3. thymine
    4. uracil

    a. 1, 2, 4
    b. 3, 4, 6
    c. 2 only
    d. 1, 2, 5
    e. 1, 2, 6

12. The backbone of the RNA molecule is composed of alternating units of

    a. ribose and phosphate.
    b. purine and pyrimidine.
    c. deoxyribose and phosphate.
    d. deoxyribose and ribose.
    e. uracil and ribose.

13. In the structure C=O how many pairs of electrons do the carbon and oxygen atoms share?

    a. one
    b. two
    c. three
    d. four
    e. none

14. Phospholipids

    a. have a polar and a nonpolar end.
    b. are found in cell membranes.
    c. function as cellular enzymes.
    d. are found only in prokaryotic cells.
    e. a and b are both correct

15. What kind of bond is formed between the oxygen and the hydrogen atoms in a water molecule?

    a. nonpolar covalent bond
    b. ionic bond
    c. polar covalent bond
    d. hydrogen bond

16. What kind of bond is formed between the oxygen of one water molecule and the hydrogen of an adjacent water molecule?

    a. nonpolar covalent bond
    b. ionic bond
    c. polar covalent bond
    d. hydrogen bond

Match the following 5 statements to the correct term in the following list:

    a. nucleic acids
    b. proteins
    c. lipids
    d. carbohydrates
    e. all of the above
    f. none of the above

17. Molecules made out of multiple (CH_2O) units.

18. Molecules that contain COOH and NH_2 groups.

19. Molecules that contain C, H and O.

20. Molecules that have alternating sugar and phosphate groups.

21. Molecules that may be oils.

Thought Questions

1. Why is water essential to life? Describe the properties of water that make it essential and relate them to the structure of the water molecule.

2. Try listing the macromolecules in order of their importance to cells. Which ones are essential?

3. Explain why it is said that life is based on the carbon atom.

Answers to Self Test Questions

1-b, 2-a, 3-c, 4-b, 5-a, 6-b, 7-a, 8-e, 9-e, 10-b, 11-d, 12-a, 13-b, 14-e, 15-c, 16-d, 17-d, 18-b, 19-d, 20-a, 21-c

Chapter 3   Microscopy and Cell Structure

Overview

Microscopes are the basic tools for the study of microscopic form and structure. This chapter introduces the various kinds of microscopes and staining techniques used by microbiologists to study both prokaryotic and eukaryotic microorganisms. The prokaryotic cell is presented with special emphasis on the cell wall, external structures, cell membrane and internal components of the cell. The eukaryotic cell is also presented with a focus on the components that are involved in microbial infection and resistance. The two kinds of cells are compared and contrasted.

Learning Objectives

After studying the material in this chapter, you should be able to:

Microscopy

1. List and describe the kinds of microscopes used in microbiology.
2. Define and contrast:
   - Magnification
   - Resolution
3. List the types of microscopes that increase contrast and explain why they are useful.
4. Describe the preparation of a smear used in staining procedures.
5. List the two kinds of differential staining procedures; give specific examples and uses of each type.
6. List the three kinds of special staining procedures; give specific examples and uses of each type.
7. Describe the Gram stain procedure and explain the function of each step.
8. Interpret a Gram stain.
9. Describe the uses of:
   - Acid fast stain
   - Capsule stain
   - Endospore stain
   - Flagella stain

Prokaryotic Cells

1. Describe the different morphological forms and arrangements of prokaryotic cells and give specific examples of each.
2. List the characteristics of prokaryotic cells.
3. Describe the structure of a prokaryotic cell and give the specific function of each of the major components.
4. Describe the fluid mosaic model of the cytoplasmic membrane.
5. Describe the major roles of the cytoplasmic membrane.
6. Describe the role of the cytoplasmic membrane in energy transformation.
7. Describe the processes by which substances are transported across cell membranes by:
   - Passive transport mechanisms
   - Active transport mechanisms
8. Define:
   - Simple diffusion
   - Osmosis
   - Osmotic pressure

9. Describe what will occur when a bacterial cell with a cell wall is place into a:
    - Hypertonic solution (higher concentration of solute outside of the cell than within of the cell)
    - Hypotonic solution (higher concentration of solute inside of the cell than outside of the cell)
10. Define:
    - Transport protein
    - Facilitated diffusion
    - Active transport
    - Proton motive force
    - ABC transport systems
    - Group translocation
11. Describe the structure of the bacterial cell wall. Differentiate between Gram-negative and Gram-positive cell walls.
12. Describe a typical bacterial chromosome.
13. Describe a plasmid and explain what it does.
14. Describe a bacterial ribosome.
15. Give the function of endospores (bacterial spores) and name the two genera in which endospores are generally found.

Eukaryotic Cells

1. List the characteristics of eukaryotic cells.
2. Explain why signaling is important in multicellular organisms.
3. Describe the plasma membrane in a eukaryotic cell.
4. Define:
    - Endocytosis
    - Exocytosis
    - Channel
    - Carrier
    - Pinocytosis
    - Phagocytosis
    - Receptor-mediated endocytosis
5. Describe the process of phagocytosis.
6. Describe how proteins are secreted.
7. Identify and give the functions of the following internal protein structures:
    - Cilia
    - Cytoskeleton
    - Flagella
    - Ribosomes
8. Describe how eukaryotic ribosomes differ in structure from prokaryotic ribosomes.
9. Describe briefly the following membrane-bound organelles and give the function of each:
    - Chloroplast
    - Endoplasmic reticulum
    - Golgi apparatus
    - Lysosome
    - Mitochondria
    - Nucleus
    - Peroxisome
10. Describe the structure of the nucleus.
11. Compare and contrast the structure and function of mitochondria and chloroplasts.
12. Describe the differences in structure and function between smooth and rough endoplasmic reticulum.

Key Concepts

1. Microscopes are the basic tools that make it possible to observe the microscopic world.
2. The functions of microscopes are to enlarge (magnify) and to make visible as separate objects (resolve) two objects that are close together.
3. The major differences between the different types of microscopes are the types of lenses and how the specimen is illuminated.
4. Staining techniques make it possible to study more clearly the structure of both prokaryotic and eukaryotic cells.
5. Prokaryotic cells are designed to carry on the basic functions of life and they have the appropriate structures to do so.
6. The structure of the bacterial cell wall is an important determinant in many of the properties of bacteria including differential staining characteristics, susceptibility to antibiotics such as penicillin and susceptibility to changes in extracellular osmotic concentrations.
7. The cytoplasmic membrane is the major determinant as to what enters and leaves a cell.
8. Prokaryotic cells have developed cellular modifications such as the glycocalyx, flagella and pili that enable them to compete successfully.
9. The bacterial chromosome consists of a single, double-stranded DNA molecule that resides in the cytoplasm of the cell rather than in a nucleus.
10. Extrachromosomal DNA called plasmids are sometimes found in prokaryotic cells and may be advantageous, but not essential, to the cell.
11. The structure of ribosomes differs in prokaryotic and eukaryotic cells making the prokaryotic ribosomes useful targets for antimicrobial agents.
12. Endospores are dormant forms within the life cycle of some bacterial species that enable the microbe to withstand difficult environmental conditions.

Summary Outline

3.1   Microscopic Techniques: The Instruments
    A.   Light Microscopes
        1.   The **bright-field microscope**: Visible light passes through the specimen
        2.   **Phase contrast microscope**: Amplifies differences in refraction
        3.   **Dark-field microscope**: Directs light toward a specimen at an angle
        4.   The **fluorescent microscope**: used to observe cells that have been stained with fluorescent dyes
        5.   The **confocal scanning laser microscope**: Used to construct a three-dimensional image of a thick structure and to provide detailed sectional views of the interior of an intact cell
    B.   **Electron Microscopes**
        1.   Use electromagnetic lenses, electrons, and phosphorus screens to produce a magnified image
            a)   **Transmission electron microscopes** (TEM) transmit electrons through a specimen that has been prepared by thin-sectioning, freeze-fracturing, or freeze-etching
            b)   **Scanning electron microscopes** scan a beam of electrons back and forth over the surface of a specimen, producing a three-dimensional effect
        1.   **Scanning probe microscopes**: Maps the bumps and valleys of a surface on an atomic scale

3.2   Microscopic Techniques: Dyes and Staining
    A. Differential stains
        1.   The Gram Stain
           a. **Gram-positive bacteria stain purple**
           b. **Gram-negative bacteria stain pink**
        2.   The **acid-fast stain**: Stains organisms such as **mycobacteria**, which do not take up stains readily; acid-fast organisms stain pink and all other organisms stain blue
    B. Special stains to observe cell structures
        1.   **Capsule stain** colors the background, allowing the capsule to stand out as a halo around an organism
        2.   **Spore stain** stains endospores
        3.   **Flagella stain** stains flagella
    C. **Fluorescent dyes and tags**: Some fluorescent dyes bind compounds that characterize all cells, others bind to compounds specific to only certain cell types
3.3   Morphology of prokaryotic cells
    A. Shapes
        1.   **Cocci**
        2.   **Rods**
        3.   **Coccobacilli**
        4.   **Vibrios**
        5.   **Spirilla**
        6.   **Spirochetes**
        7.   **Pleomorphic** bacteria have variable shapes
    B. Groupings: Cells adhering to one another following division form a characteristic arrangement that depends on the plane in which the bacteria divide
    C. Multicellular associations
        1.   Associations containing multiple cells, such as myxobacteria
        2.   **Biofilms** often alter their activities when a critical number of cells are present
3.4   The Prokaryotic Cell: The cytoplasmic membrane
    A. Structure and chemistry of the cytoplasmic membrane
        1.   **Phospholipid bilayer** embedded with a variety of different **proteins**
        2.   **Differential barrier** between the cell and the surrounding environment
        3.   Membrane **proteins function in transport** or provide a mechanism by which cells can **sense** and **adjust** to their surroundings
    B. Permeability of the cytoplasmic membrane
        1.   **Selectively permeable**
        2.   Inflow of water into the cell exerts more osmotic pressure on the cytoplasmic membrane than it can generally withstand
    C. The role of the cytoplasmic membrane is involved in **energy generation**: Electron transport chain within the membrane expels protons, generating an electrochemical gradient, which contains a form of energy called proton motive force
3.5   The Prokaryotic Cell: Directed movement of materials across the cytoplasmic membrane
    A. **Transport systems**
        1.   **Facilitated diffusion**: Moves impermeable compounds from one side of the membrane to the other by exploiting the concentration gradient
        2.   **Active transport** mechanisms use energy to accumulate compounds against a concentration gradient
        3.   Members of the major facilitator superfamily use **proton motive force** for energy
        4.   **ABC transport** systems require ATP for energy
        5.   **Group translocation** chemically modifies a molecule during its passage through the cytoplasmic membrane

B. **Secretion**: The general secretory pathway is the primary mechanism used to secrete proteins

3.6   The Prokaryotic Cell: Cell Wall

    A. **Peptidoglycan**

        1. Found only in the bacteria and provides rigidity to the cell wall

        2. Composed of peptidoglycan which contains alternating subunits of *N*-**acetylmuramic acid (NAM)** and *N*-**acetylglucosamine (NAG)** interconnected via the tetrapeptide chains on NAM

    B. The **Gram-positive cell wall**

        1. Relatively **thick layer of peptidoglycan**

        2. **Teichoic acids** and **lipoteichoic acids** stick out of the peptidoglycan molecule

    C. The Gram-negative cell wall

        1. **Thin layer of peptidoglycan** sandwiched between the cytoplasmic membrane and an outer membrane

        2. **Periplasm** contains a variety of proteins

        3. The outer membrane contains lipopolysaccharides. **The Lipid A portion** of the lipopolysaccharide molecule is toxic, which is why LPS is called **endotoxin**

        4. **Porins** form small channels that permit small molecules to pass through the outer membrane

    D. Antibacterial compounds that target peptidoglycan

        1. Penicillin binds to proteins involved in cell wall synthesis

        2. Lysozyme breaks the bond that links alternating NAG and NAM molecules

    E. Characteristics of bacteria that lack a cell wall

        1. Because **mycoplasmas** do not have a cell wall, they are extremely variable in shape and are not effected by lysozyme or penicillin

    F. Cell walls of the Domain Archaea have a greater variety than those of the Domain Bacteria

3.7   The Prokaryotic Cell: Surface layers external to the cell wall

        1. **Glycocalyx**: Enable bacteria to adhere to surfaces; some capsules allow disease-causing microorganisms to thwart the innate defense system

            a) **Capsule**: A distinct and gelatinous layer made of polysaccharide

            b) **Slime layer**: Diffuse and irregular layer of polysaccharide

3.8.   The Prokaryotic Cell: Filamentous protein appendages

        1. **Flagella**

            a) Long protein structures responsible for most types of bacterial motility

            b) **Chemotaxis** is the directed movement toward an attractant or away from a repellant

        2. Pili

            a) Many types of **pili (fimbriae)** enable attachment of cells to specific surfaces

            b) Some pili play a role in specific types of motility

            c) **Sex pili** are involved in conjugation, which enables DNA to be transferred from one cell to another

3.9.   The Prokaryotic Cell: Internal structures

    A. **Chromosome**

        1. The chromosome of prokaryotes resides in the nucleoid

        2. The typical chromosome is a **single, double-stranded DNA molecule** that contains all the genetic information required by a cell

    B. **Plasmids**

        1. Plasmids are **circular, double-stranded DNA** molecules that typically encode genetic information that may be advantageous, but not required by the cell

        2. Populations of cells can gain and lose plasmids, depending on the relative advantages

    C. **Ribosomes**

        1. Ribosomes facilitate the joining of amino acids

        2. The **70S bacterial ribosome** is composed of a 50S and a 30S subunit

    D. **Storage granules**: Dense accumulations of high molecular weight polymers, which are synthesized from a nutrient that a cell has in relative excess

    E. **Gas vesicles** are gas-permeable, water-impermeable rigid structures that provide buoyancy to aquatic cells

    F. **Endospores**

        1. Dormant stage produced by members of the **genera *Bacillus*** and ***Clostridium***; they can germinate to become a vegetative cell

        2. **Resistant** to conditions such as heat, desiccation, toxic chemicals, and UV irradiation

        3. **Sporulation** is an eight-hour process initiated when cells are grown under nutrient-limiting conditions

        4. **Germination** is the process by which an endospore leaves its dormant state

3.10 The Eukaryotic Cell: The **plasma membrane**

    A. **Phospholipid bilayer** embedded with **proteins**

    B. Proteins are involved in **transport, structural integrity** and **signaling**

3.11 The Eukaryotic Cell: Transfer of molecules across the plasma membrane

    A. **Transport proteins**

        1. **Channels** are pores in the membrane so small that only specific ions can pass. These channels are gated

        2. Cells of multicellular organisms often take up nutrients by facilitated diffusion because the nutrient concentration of their surrounding environment can be controlled

        3. **Carriers** involved in active transport include members of the major facilitator superfamily and ABC transporters

    B. **Endocytosis** and **exocytosis**

        1. **Receptor-mediated endocytosis**

        2. Protozoa and phagocytes take up bacteria and debris through the process of **phagocytosis**

        3. **Exocytosis** expels products and it the reverse of endocytosis

    C. **Secretion**

        1. Proteins are made by ribosomes bound to the endoplasmic reticulum

        2. The proteins are threaded through the membrane and into the lumen of the endoplasmic reticulum

3.12 The Eukaryotic Cell: Protein structures within the cytoplasm

    A. **Ribosomes**: The 80S eukaryotic ribosome is composed of 60S and 40S subunits

    B. **Cytoskeleton**

        1. **Microtubules** are the thickest of the cytoskeleton structures and are long hollow cylinders

        2. **Microfilaments** allow the cytoplasm to move and are composed of actin

        3. **Intermediate filaments** strengthen the cell mechanically

    C. Flagella and cilia

        1. **Flagella** and **cilia** are composed of microtubules in a 9+2 arrangement

        2. Flagella propel a cell or pull the cell forward

        3. Cilia often cover the surface of a cell and move in synchrony to either propel a cell or move material along a stationary cell

3.13 The Eukaryotic Cell: Membrane-bound organelles

    A. The **nucleus**

        1. The nucleus **contains DNA** and is the predominant distinguishing feature of eukaryotes

        2. Two membranes compose the **nuclear envelope**

        3. **Nuclear pores** allow large molecules to be transported in and out of the nucleus

        4. The **nucleolus** is where ribosomal RNAs are synthesized and, along with ribosomal proteins, are assembled into ribosomal subunits

B. Mitochondria and chloroplasts
   1. Contained within the inner membrane of mitochondria and chloroplasts are the proteins of the electron transport chain and proteins that use proton motive force to generate ATP
   2. **Mitochondria** use the energy released during the degradation of organic compound to generate ATP
   3. **Chloroplasts** contain chlorophyll, which captures the energy of sunlight; this is then used to synthesize ATP
C. **Endoplasmic reticulum**
   1. **Rough** endoplasmic reticulum
      a) Lined with ribosomes
      b) Serves as the site where proteins that are not located in the cytoplasm are synthesized
   2. **Smooth** endoplasmic reticulum
      a) Lipids are synthesized and degraded, and calcium is stored
D. **Golgi apparatus**: Modifies and sorts molecules synthesized in the ER
E. **Lysosomes**: Structures within which digestion takes place
F. **Peroxisomes**: Organelles in which oxygen is used to oxidize certain substances

Terms You Should Know

Microscopy and Cell Morphology

Bacillus
Binary fission
Capsule
Coccobacillus
Coccus
Compound microscope
Condenser lens
Contrast
Decolorizing agent
Diplococcus
Electron photomicrograph
Endospore
Fluorescent dyes
Immunofluorescence
Mordant
Objective lens
Ocular lens
Pleomorphic
Refraction
Resolution
Spirillum
Spirochete

Prokaryotic Cell

Active transport
Capsule
Cell wall
Chemotaxis
Chromosome

Cytoplasm
Cytoplasmic membrane
Electrochemical gradient
Endospore
Endotoxin
Facilitated diffusion
Flagella
Fluid mosaic model
Gas vesicles
Germination
Glycocalyx
Group translocation
Lipid A
Lipopolysaccharide layer (LPS)
Major facilitator superfamily (MFS)
N- acetylglucosamine (NAG)
N-acetylmuramic acid (NAM)
Nucleoid
O polysaccharide
Osmosis
Osmotic pressure
Passive transport
Peptidoglycan
Pili (fimbriae)
Plasmid
Porins
Proton motive force
Pseudopeptidoglycan
Ribosome
Selectively permeable
Sex pili

Simple diffusion
Slime layer
Sporulation
Storage granules
Teichoic acids
Transport proteins (permeases or carriers)

Eukaryotic cell

Chloroplast
Chromatin
Cilia
Cytoskeleton
Diploid
Endocytosis
Endoplasmic reticulum
Exocytosis
Golgi apparatus
Histone
Lysosome
Meiosis
Mitochondria
Mitosis
Nucleolus
Nucleus
Organelle
Peroxisome
Phagocytosis
Photosynthesis
Pinocytosis
Receptor-mediated endocytosis

Learning Activities

1. On the figure of the microscope label the following:

   Eyepiece, stage, condenser, iris diaphragm, objective lens and light source

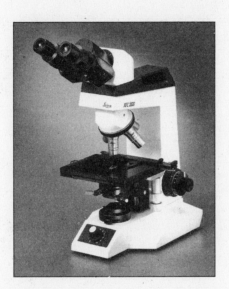

2. Describe the following techniques:
   - smear preparation _____
     _____

   - aseptic transfer _____
     _____

3. The following is a list of the kinds of microscopes used in microbiology. Briefly give the advantage of each.

Type of Microscope	Advantage
1. Bright-field	
2. Dark-field	
3. Phase-Contrast	
4. Differential Interference Contrast	
5. Fluorescence	
6. Confocal	
7. Scanning Electron	
8. Transmission Electron	
9. Scanning Tunneling	
10. Atomic Force	

4. Give the function of each of the listed staining procedures.

Type of Stain	Function of stain
1. Simple stain	
2. Gram stain	
3. Acid-fast stain	
4. Negative stain	
5. Endospore stain	
6. Flagella stain	

5. Complete the following table on the Gram stain.

Step	Reagent	Function
Primary Stain		
Mordant		
Decolorizer		
Counterstain		

6. On the drawing of the prokaryotic cell label the following:

    Glycocalyx, cell wall, cytoplasmic membrane, pili, chromosome, flagellum, nucleoid, storage granule, ribosomes

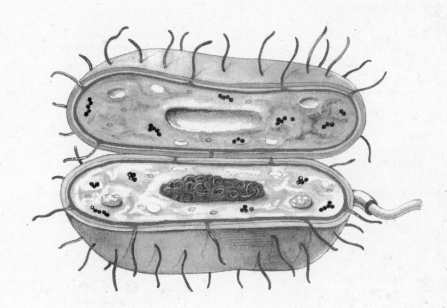

7. Compare the characteristics of prokaryotic and eukaryotic cells.

Characteristic	Prokaryotic Cell	Eukaryotic Cell
Location of DNA		
Organelles		
Histones		
Peptidoglycan		
Reproduction		

8. Give specific examples of organisms that have:

Type	Example
Gram-positive cell walls	
Gram-negative cell walls	
No cell wall	

9. Give an example for each of the following morphological forms and arrangements of prokaryotic cells.

Shape and arrangement	Example
Diplococcus	
Bacillus (single)	
Coccobacillus	
Vibrios	
Spirilla	
Spirochete	
Pleomorphic rod	

10. List the major differences between a Gram-negative and Gram-positive cell wall.

Gram-Positive Cell Wall	Gram-Negative Cell Wall

11. For the list of cell types, indicate the type of cell wall each has (including none or incomplete) and the major component of the wall.

Cell Type	Cell wall	Major component
Gram-positive bacteria		
Gram-negative bacteria		
Archaea		
Mycoplasmas		
Fungi		
Plant		
Animal		

12. Describe the structures found in a prokaryotic cell and give the function of each.

Structure	Description	Function
Glycocalyx		
Flagella		
Pili		
Gram-positive cell wall		
Gram-negative cell wall		
Cytoplasmic membrane		
Cytoplasm		
Chromosome		
Plasmid		
Ribosomes		
Storage granules		
Endospores		

13. On the drawing of the Gram-positive cell wall label the following:

Peptidoglycan, *N*-acetylglucosamine, *N*-acetylmuramic acid, teichoic acid, cytoplasmic membrane

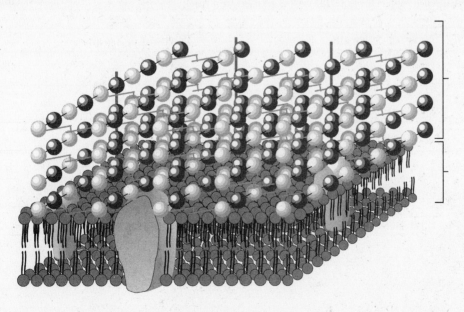

14. On the drawing of the Gram-negative cell wall label the following:

    Lipopolysaccharide, peptidoglycan, porin protein, outer membrane, periplasm, cytoplasmic membrane

15. Describe what will occur when a bacterial cell is placed in a (an):

Condition	With cell wall	Without cell wall
Hypertonic solution		
Hypotonic solution		
Isotonic solution		

16. List the components of the cytoplasmic membrane.

17. On the drawings of the eukaryotic cell label the following:

Plasma membrane, nucleus, nucleolus, mitochondria, endoplasmic reticulum, Golgi apparatus, chloroplast, cilia, lysosomes, ribosomes

Typical animal cell

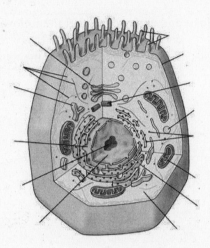

Typical plant cell

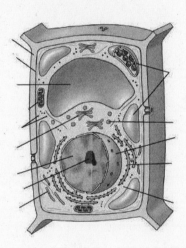

18. Give the function of endospores (bacterial spores).

_____

_____

_____

19. List the genera in which endospores are generally found.

_____

_____

_____

20. List two organelles in eucaryotic cells that resemble a prokaryotic cell.

_____

_____

21. On the drawing illustrating endocytosis and exocytosis label the following:

Pinocytosis, phagocytosis, exocytosis, endocytic vesicle, endosome, lysosome, phagosome, phagolysosome, pseudopod

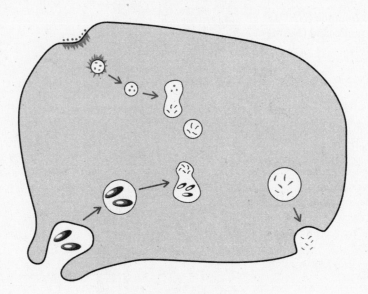

Instruments and Techniques to Know

Instruments

Describe the use of each of the following kinds of microscopes.

Bright field microscope
Confocal microscope
Dark field microscope
Fluorescence microscope
Interference microscope
Phase contrast microscope
Scanning electron microscope
Scanning probe microscopes
    Atomic force
    Scanning tunneling
Transmission electron microscope

Techniques

Acid-fast stain
Capsule stain
Differential staining
Endospore stain
Flagella stain
Freeze-etching
Freeze-fracturing
Gram stain
Negative staining
Simple staining
Smear preparation
Thin-sectioning

Self Test

1. The cell wall of Gram-positive bacteria is mainly composed of

    a. cellulose.
    b. protein.
    c. phospholipids.
    d. peptidoglycan.
    e. starch.

2. Peptidoglycan is composed of

    a. N-acetylglucosamine (NAG)
    b. N-acetylmuramic acid (NAM)
    c. Teichoic acids
    d. only a and b
    e. a, b and c

3. The cell wall of Gram-negative bacteria differs from that of Gram-positive bacteria in that Gram-negative bacteria

    a. have more peptidoglycan.
    b. have teichoic acids.
    c. have an outer membrane composed of lipopolysaccharides and proteins.
    d. are very resistant to physical disruption.

4. Which of the following structures is essential for all cells?

    a. Pili
    b. Flagella
    c. Cell wall
    d. Cytoplasmic membrane
    e. Capsule

5. Ribosomes are organelles that are found in both prokaryotic and eukaryotic cells. What is the function of ribosomes?

    a. Synthesis of protein
    b. Replication of the cell
    c. Transformation of energy
    d. Transportation of substances across the cell membrane
    e. None of the above

6. The group of bacteria that do not have cell walls are the

    a. Archaea.
    b. Mycoplasma.
    c. Bacteria.
    d. Mycobacteria.

7. Which of the following does not have a membrane separating their nuclear material from their cytoplasm?

    a. prokaryotes
    b. eukaryotes
    c. mycoplasma
    d. yeasts

8. Which of the following cellular structures helps protect organisms from phagocytosis?

    a. pili
    b. flagella
    c. endospore
    d. capsule
    e. ribosomes

9. Endotoxins are

    a. found only in Gram-positive cell walls.
    b. found only in Gram-negative cell walls.
    c. composed of lipopolysaccharides.
    d. found only in mycoplasma.
    e. both b and c

10. Endospores

    a. are formed by the process of sporulation.
    b. are produced by bacteria of the genera *Bacillus* and *Clostridium*.
    c. are dormant forms produced under nutrient-limiting conditions.
    d. are extraordinarily resistant to heat and desiccation.
    e. all of the above.

Thought Questions

1. List four essential functions of cells.

2. To control microbial growth, microbial cells must be targeted by antimicrobial agents. List three ways in which microbial cells can be targeted without harming human cells.

Answers to Self Test Questions

1-d, 2-d, 3-d, 4-d, 5-a, 6-b, 7-a, 8-d, 9-b, 10-e

Chapter 4   Dynamics of Prokaryotic Growth

Overview

This chapter describes how bacteria are cultivated in the laboratory. Bacterial growth is defined and described. Methods to detect and measure bacterial growth are presented. Environmental factors that influence microbial growth are identified and related to specific groups of bacteria. Nutritional factors are presented. The growth curve and its component phases are discussed. The interactions of mixed microbial communities in natural environments and biofilms are discussed.

Learning Objectives

After studying the material in this chapter, you should be able to:

1. Define pure culture and explain its significance.
2. Define colony.
3. Describe the streak plate method for the isolation of bacteria.
4. Describe how bacteria can be preserved for extended periods of time.
5. Define binary fission.
6. Explain generation or doubling time.
7. List and describe the chemical and physical conditions necessary for bacterial growth.
8. Classify bacteria on the basis of temperature preference and tolerance.
9. Classify bacteria on the basis of oxygen utilization and tolerance.
10. Describe the roles of superoxide dismutase and catalase in oxygen utilization and/or tolerance.
11. Give the categories based on the oxygen requirements of bacteria that have
    * Superoxide dismutase
    * Catalase
12. List the nutritional factors that influence microbial growth.
13. Define growth factors.
14. List and describe the four nutritional groups of organisms based on energy source and carbon sources. Indicate what kind of organisms are in each group.
15. Explain how bacterial growth requirements are provided in laboratory cultures.
16. List the basic types of media used in the bacteriological laboratory.
17. Differentiate between complex and chemically defined media.
18. Define, describe and give examples of the following kinds of media:
    * Selective media
    * Differential media
19. Describe the use of the following equipment in the bacteriological laboratory:
    * Candle jar
    * Carbon dioxide incubator
    * Anaerobe jar
    * Anaerobe incubator
20. Describe how direct cell counts are done.
21. Describe how viable cell counts are done.

22. Describe the most probable number (MPN) method.
23. Explain how biomass is measured.
24. Explain how cell products can be used to measure bacterial growth.
25. Describe logarithmic growth.
26. Draw and describe the bacterial growth curve identifying the events occurring in each section.
27. Explain how a bacterial culture can be sustained for a long period of time of time.
28. Define biofilm and explain its importance.

Key Concepts

1. Pure culture is a population of cells descended from a single cell.
2. Colony is a mass of bacterial cells that originated from one cell.
3. Isolation of bacteria refers to methods by which bacteria can be separated into pure cultures.
4. Binary fission is the type of asexual reproduction by which a bacterial cell splits and forms two daughter cells.
5. Generation or doubling time is the amount of time that is necessary for a population of bacteria to undergo one round of binary fission thereby doubling the number of bacteria present.
6. Bacteria have both physical and chemical requirements for growth, which may include temperature, oxygen levels, pH, water, required elements, organic growth factors and carbon and energy sources.
7. Microorganisms can be separated into groups based on temperature, oxygen levels and pH requirements.
8. Organisms can be separated into four groups based on their carbon and energy sources.
9. Growth in the laboratory requires that the needs of the bacteria be provided for in media.
10. Media is the material that provides nutrition for bacterial growth in culture.
11. Measurement of microbial growth can be accomplished by direct microscopic counts, cell counting instruments, plate counts, membrane filtration, most probable number method, turbidity, total weight of a culture and the presence of certain cell constituents.
12. Spectrophotometers, Coulter counters and flow cytometers are instruments that can be used to measure microbial growth. In a closed system a population of bacterial will follow growth curve in which there is a lag phase, a logarithmic growth phase followed by a stationary phase and a logarithmic decline phase.
13. Bacteria may be kept in a continuous growth stage of growth by supplying nutrients using a chemostat.
14. Bacterial growth in natural environments is generally more dynamic than under artificial conditions resembling that of a continuous culture.
15. Biofilms are communities of bacteria with characteristic structure containing open channels through which nutrients and waste products may pass.

Summary Outline

4.1 Obtaining a **pure culture**
  A. About one-tenth of one percent of bacteria can be cultured in the laboratory.
  B. Cultivating bacteria on a solid medium
    1. A single bacterial cell will multiply to form a visible colony.
    2. **Agar** is used to solidify nutrient-containing broth.
  C. The **streak plate method** is used to isolate bacteria in order to obtain a pure culture.

D. Maintaining stock cultures
   1. **Stock cultures** can be used as an **inoculum** in later experiments.
   2. **Stock cultures** can be **stored on an agar slant in the refrigerator**, **frozen in a glycerol solution** or **lyophilized**.

4.2 Principles of bacterial growth
   A. Most bacteria multiply by **binary fission**.
   B. **Microbial growth** is an **increase in the number of cells** in a population.
   C. The time required for a population to double in number is the **generation time**.

4.3 **Environmental factors** that influence microbial growth
   A. **Temperature requirements**
      1. **Psychrophiles** have an optimum between -5°C and 15°C.
      2. **Psychrotrophs** have an optimum between 20°C and 30°C
      3. **Mesophiles** have an optimum between 25°C and 45°C.
      4. **Thermophiles** have an optimum between 45°C and 70°C.
      5. **Hyperthermophiles** have an optimum between 70°C and 110°C.
      6. Storage of foods at refrigeration temperatures retards spoilage because it limits the growth of mesophiles.
      7. Some microorganisms can inhabit certain parts of the body but not others because of temperature differences.
   B. **Oxygen requirements**
      1. **Obligate anaerobes** cannot multiply if oxygen is present.
      2. **Facultative anaerobes** can multiply if oxygen is present but can also grow without it.
      3. **Microaerophiles** require small amounts of oxygen but higher concentrations are inhibitory.
      4. **Aerotolerant anaerobes** are indifferent to oxygen.
      5. Oxygen can be converted to **superoxide** and **hydrogen peroxide**, both of which are toxic. **Superoxide dismutase** and **catalase** can break these down.
   C. **PH**
      1. Most bacteria live within the pH range of 5 to 8.
      2. **Acidophiles** grow optimally at a pH below 5.5.
      3. **Alkaliphiles** grow optimally at a pH above 8.5.
   D. **Water** availability
      1. All microorganisms require water for growth.
      2. If the solute concentration is higher in the medium than in the cell, water diffuses out of the cell, causing plasmolysis.
      3. **Halophiles** have adapted to live in high salt environments.

4.4 **Nutritional factors** that influence microbial growth
   A. Required elements
      1. The major elements make up cell constituents and include **carbon, nitrogen, sulfur** and **phosphorus**.
      2. **Heterotrophs** use organic carbon.
      3. **Autotrophs** fix CO_2.
      4. **Trace elements** are required in very minute amounts.
   B. **Growth factors** are cell constituents such as amino acids and vitamins that the cell cannot synthesize.
      1. Organisms derive energy either from sunlight or from the oxidation of chemical compounds.
      2. Nutritional diversity

C. Prokaryotes use diverse sources of carbon and energy.
 1. **Photoautotrophs** use the energy of sunlight and the carbon in the atmosphere to make organic compounds.
 2. **Chemolithoautotrophs** use inorganic compounds for energy and derive their carbon from CO_2.
 3. **Photoheterotrophs** use the energy of sunlight and derive their carbon from organic compounds.
 4. **Chemoorganoheterotrophs** use organic compounds for energy and as a carbon source.

4.5 Cultivating prokaryotes in the laboratory
 A. General categories of culture media
  1. **Complex medium** contains a variety of ingredients such as peptones and extracts. (Examples: nutrient agar, blood agar and chocolate agar)
  2. A **chemically defined medium** is composed of precise mixtures of pure chemicals; an example is glucose-salts medium.
 B. Special types of culture media
  1. A **selective medium** inhibits organisms other than the one being sought (Examples: **Thayer Martin agar** and **MacConkey agar**)
  2. A **differential medium** contains a substance that certain bacteria change in a recognizable way (Examples: **Blood agar** and **MacConkey agar**)
 C. Providing appropriate **atmospheric conditions**
  1. A **candle jar** provides **increased CO_2**, which enhances the growth of many medically important bacteria.
  2. **Microaerophilic bacteria** are incubated in a gas-tight jar along with atmospheric oxygen to form water.
  3. **Anaerobes** may be cultivated in either an **anaerobe jar** or a medium that incorporates a reducing agent.
  4. An enclosed chamber that maintains anaerobic conditions can also be used.
  5. **Enrichment cultures** provide conditions in a broth that enhance the growth of one particular organism in a mixed population.

4.6 Methods to detect and measure bacterial growth
 A. Direct cell counts generally do not distinguish between living and dead cells.
  1. **Direct microscopic count**
  2. The **Coulter counter** and a **flow cytometer** count cells as they pass through a minute aperture.
  3. **Viable cell counts**
  4. **Plate counts** are based on the fact that an isolated cell will form a single colony.
  5. **Membrane filtration** concentrates bacteria by filtration.
  6. The **most probable number (MPN) method** is a statistical assay based on the theory of probability and is used to estimate cell numbers.
 B. **Measuring biomass**
  1. **Turbidity** of a culture is a rapid measurement that can be correlated to the number of cells; a spectrophotometer is used to measure turbidity.
  2. **Wet weight** and **dry weight** are proportional to the number of cells in a culture.
  3. The **quantity of a cell constituent** such as nitrogen can be used to calculate biomass.
 C. **Measuring cell products**
  1. **pH indicators** can be used to monitor acid production.
  2. **Gas production** can be detected by pH changes or by using an inverted tube in culture media to trap gas.
  3. **ATP** is detected by employing luciferase.

4.7 Bacterial growth in laboratory conditions
   A. Bacterial growth follows a **growth curve** when they are grown in a closed system.
      1. **Lag**—number of cells does not increase
      2. **Log**—cells divide at a constant rate
      3. **Stationary**—a required nutrient is used up, oxygen is in short supply, or toxic metabolites accumulate
      4. **Death**—number of viable cells in the population decreases
      5. **Prolonged decline**- gradual decrease in the number of viable cells in the population over a long period of time
   B. **Colony growth**: The position of a single cell within a colony markedly determines its environment; cells on the edge may be in log phase whereas those in the center may be in the death phase.
   C. **Continuous cultures**: Bacteria can be maintained in a state of continuous exponential growth by using a chemostat.
4.8 Bacterial growth in nature
   D. **Mixed populations**: Bacteria often grow in close associations with other kinds of organisms; the metabolic activities of one organism may facilitate the growth of another organism.
   E. **Biofilms**: Bacteria may live suspended in an aqueous environment but many attach to surfaces and live as a biofilm, a polysaccharide-encased community.

Terms You Should Know

Aerotolerant anaerobes
Agar
Alpha-hemolysis
Aseptic techniques
Autotrophs
Beta-hemolysis
Binary fission
Biofilm
Blood agar
Capnophiles
Carbon fixation
Catalase
Chemically defined media
Chemolithoautotrophs
Chemoorganoheterotrophs
Chemotrophs
Chocolate agar
Complex media
Death or decline phase
Differential media

Durham tube
Enrichment culture
Exponential or log phase
Facultative anaerobes
Growth curve
Growth factors
Hemolysin
Heterotrophs
Hyperthermophiles
Lag phase
Lyophilization
MacConkey agar
Mesophiles
Microaerophiles
Nitrogen fixation
Nutrient agar
Nutrient broth
Obligate aerobes
Obligate anaerobes
Petri dish

pH
pH indicator
Phase of prolonged decline
Photoautotrophs
Photoheterotrophs
Phototrophs
Plasmolysis
Psychrophiles
Psychotrophs
Pure culture
Selective media
Stationary phase
Sterile
Superoxide
Superoxide dismutase
Thayer-Martin agar
Thermophiles
Trace elements
Turbidity

Learning Activities

1. Complete the following table.

Classification by oxygen utilization and tolerance	Use/requirement/tolerance of oxygen	Enzymes produced
Obligate aerobes		
Facultative anaerobes		
Obligate anaerobes		
Aerotolerant anaerobes		
Microaerophiles		

2. List the five groups of bacteria based on temperature preference and give the approximate temperature range of each.

Group of bacteria	Temperature range

3. On the drawings label the following:

    Obligate aerobe, facultative anaerobe, obligate anaerobe, microaerophile, aerotolerant

    Indicate which types of bacteria have: Catalase, superoxide dismutase

Table 4.3 Oxygen (O$_2$) Requirements of Prokaryotes

4. Give the functions of the following enzymes:

Enzyme	Function
Superoxide dismutase	
Catalase	
Peroxidase	

5. Explain how bacterial growth requirements are provided in laboratory cultures.

_____

_____

_____

6. Define or identify the following kinds of media.

Type of media	Definition
Defined	
Complex	
Selective	
Differential	

7. Give the function and an example of the following kinds of media.

Type of media	Function	Example
Defined		
Complex		
Selective		
Differential		

8. Describe the use of the following equipment in the bacteriological laboratory:

Equipment	Use or function
Anaerobe jar	
Anaerobe chamber	
Candle jar	
Carbon dioxide incubators	

9. Define the following:

Term	Definition
Pure culture	
Colony	

10. Describe the use of the following instruments and techniques.

Instrument/Technique	Use
Coulter counter	
Flow cytometer	
Spectrophotometer	
Most probable number method	
Pour plate	
Spread plate	
Membrane filtration	

11. Using the following data and chart give the most probable number for each.

Tubes of nutrient media (Sets of five)	Number of positive tubes per set			
10 ml of inoculum	4	5	4	5
1 ml of inoculum	2	0	3	1
0.1 ml of inoculum	0	1	0	0
MPN index/100 ml				

Number of positive tubes in set of five	MPN index/100 ml
4-0-0	13
4-0-1	17
4-1-0	17
4-1-1	21
4-1-2	26
4-2-0	22
4-2-1	26
4-3-0	27
4-3-1	33
4-4-0	34
5-0-0	23
5-0-1	30
5-0-2	40
5-1-0	30

12. Describe what occurs in each of the five phases of the bacterial growth curve.

Growth Phase	What occurs
Lag phase	
Exponential or logarithmic growth phase	
Stationary phase	
Death phase	
Phase of prolonged decline	

13. Draw and label the bacterial growth curve.

14. If you have one bacterium with a generation time of 20 minutes how many would you have after:

Time	Approximate number of bacteria
0	1
20 minutes	
1 hour	
3 hours	
10 hours	
24 hours	
48 hours	

15. Describe the streak plate method for the isolation of bacteria.

_____

_____

16. Complete the following table.

Nutritional Type	Source of energy	Source of carbon
Photoautotrophs		
Photoheterotrophs		
Chemolithoautotrophs		
Chemoorganoheterotrophs		

17. From the drawing illustrating serial dilutions answer the following:

How much culture would you transfer from
       1.   Tube 1 to Tube 2?
       2.   Tube 2 to Tube 3?
       3.   Tube 3 to Tube 4?

What is the dilution of
       1.   Tube 2?
       2.   Tube 3?
       3.   Tube 4?

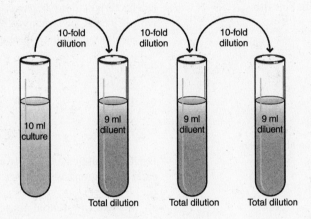

18. On the drawing illustrating temperature requirements for growth of bacteria label:

Psychrophiles, psychrotrophs, mesophiles, thermophiles, hyperthermophiles

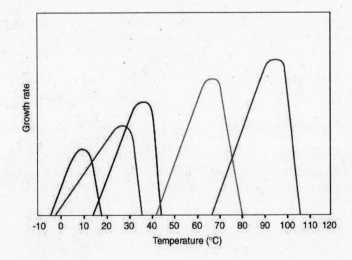

Self Test

1. Which of the following statement(s) is/are true about bacterial growth?

    a. Bacteria grow by enlarging in size.
    b. Bacteria grow by increasing the number of organisms in the population.
    c. Bacteria reproduce by binary fission.
    d. Bacteria reproduce by mitosis.
    e. Both B and C are correct.

2. Which of the following is essential for the growth of all bacteria?

    a. vitamins
    b. oxygen
    c. water
    d. glucose
    e. phospholipids

3. Bacteria that can grow only in the absence of molecular oxygen are called

    a. obligate aerobes.
    b. obligate anaerobes.
    c. facultative aerobes.
    d. facultative anaerobes.
    e. halophiles.

4. Bacteria isolated from Antarctic marine fish are probably

    a. psychrophiles.
    b. mesophiles.
    c. halophiles.
    d. thermophiles.
    e. both a and c

5. The best definition of generation time of a bacterial population is the

    a. time it takes for a bacterial population to double.
    b. time it takes for the lag phase.
    c. length of the exponential phase.
    d. time it takes for nuclear division.
    e. maximum rate of doubling.

6. In which of the following phases of the bacterial growth curve are more bacteria being produced than are dying?

    a. lag phase
    b. log phase
    c. stationary phase
    d. death phase
    e. both b and d

7. If a bacterial culture began with 4 cells and ended with 128 cells, how many generations did the bacterial population go through?

    a. 64
    b. 32
    c. 6
    d. 5
    e. 4

8. Bacterial cultures eventually stop growing and enter the stationary phase because they

    a. deplete an essential nutrient.
    b. accumulate toxic products.
    c. become too crowded.
    d. both a and b
    e. both a and c

9. The approximate optimal pH for the growth of most bacteria:

    a. 1
    b. 5
    c. 7
    d. 9
    e. 14

10. A facultative anaerobic organism

    a. doesn't use oxygen, but is not killed by it.
    b. is killed by oxygen.
    c. grows in the presence or absence of oxygen.
    d. requires oxygen, but in lower amount than is present in air.
    e. prefers to grow without oxygen.

Thought Questions

1. Differentiate between bacterial growth in the laboratory and in natural environments.

2. Explain the significance of biofilms.

Answers to Self Test Questions

1-e, 2-c, 3-b, 4-e, 5-a, 6-b, 7-d, 8-d, 9-c, 10-c

Chapter 5   Control of Microbial Growth

Overview

Control of microbial growth is essential to preventing infections, limiting the spread of disease and preserving foodstuffs and goods. In this chapter the physical and chemical methods for the control of microbial growth are presented. Situational considerations as well as the various techniques are discussed. Characteristics of specific chemicals are presented. Selection and applications of the appropriate physical or chemical methods based on the type of microorganism, the numbers of microorganisms present, environmental conditions and potential risk of infection are discussed.

Learning Objectives

After studying the material in this chapter, you should be able to:

1. List the conditions that influence the selection of a particular antimicrobial procedure.
2. List and describe the physical methods of microbial control and give their applications.
3. Describe the general action of microbial agents.
4. Differentiate between antiseptics and disinfectants.
5. Describe the factors that should be considered in selection of an appropriate antimicrobial chemical.
6. List and describe the chemical methods of microbial control (antiseptics/disinfectants) and give their applications.
7. Describe how food products and other products can be preserved by preventing the growth of microbes.

Key Concepts

1. Sterilization is the process by which all microorganisms are killed.
2. Disinfection is the process in which the number of microbes is reduced to a level where they are no longer a problem.
3. Both physical and chemical methods can be used to sterilize or disinfect.
4. Physical methods of control include both moist and dry heat treatment, irradiation, filtration and mechanical removal.
5. Methods of control used depend on the situation and the degree of control required.
6. The type of microbe, the numbers present, environmental conditions and the potential risk of infection must be considered in selecting the appropriate method of sterilization or disinfection.
7. Moist heat such as boiling destroys vegetative bacterial cells and many viruses.
8. Pasteurization does not kill all microorganisms present, but significantly reduces the numbers of heat-sensitive organisms.
9. Autoclaves use live steam under pressure to microbes, viruses and endospores.
10. Dry heat also kills microorganisms, but requires a significantly greater length of time.
11. The canning process is specifically designed to destroy the endospores of *Clostridium botulinum*.
12. Antimicrobial chemicals, which can be used to disinfect and, under some circumstances, sterilize, are less reliable than heat.

13. Bacterial endospores of *Bacillus* and *Clostridium*, *Mycobacterium* species, *Pseudomonas* species and naked viruses are resistant to antimicrobial treatment.
14. Microorganisms and viruses can be removed from liquids and air by filtration.
15. Gamma irradiation can be used to sterilize materials and to decrease the number of microorganisms in foods. Ultraviolet light is not very penetrating, but can be used to disinfect surfaces.
16. Preservation of foodstuffs to delay spoilage may be accomplished by slowing or stopping the growth of the microbes.

Summary Outline

5.1. Approaches to Control: The methods used to destroy or remove microorganisms and viruses can be:
    A.   **Physical** such as heat treatment, irradiation and filtration or **Chemical**
    B.   Principles of control
        1.   **Sterilization** destroys all microorganisms and viruses.
        2.   **Disinfection** eliminates most disease-causing bacteria or viruses.
        3.   **Disinfectants** are chemicals used for disinfecting inanimate objects
        4.   **Antiseptics** are chemicals formulated for use on skin.
        5.   **Pasteurization** uses heat treatment to reduce the number of spoilage organisms or kill disease-causing microbes.
    C.   **Situational considerations**
        1.   Hospitals must be scrupulous in controlling microorganisms because of the danger of **nosocomial infections**.
        2.   Microbiology laboratories must use aseptic technique to avoid contaminating cultures with extraneous microbes and to protect workers and the environment from contamination.
        3.   Foods and other perishable products retain their quality and safety when the growth of contaminating microorganisms is prevented.
5.2   **Selection** of an antimicrobial procedure
    A.   Type of microorganism
        1.   **Type** of microbial population present.
        2.   The **endospores** of **Bacillus** and **Clostridium** are most resistant.
        3.   The **waxy cell wall** of **mycobacteria** makes them resistant.
        4.   *Pseudomonas* are common environmental organisms are very resistant.
        5.   **Viruses** that **lack a lipid envelope** are more resistant to disinfectants than are enveloped viruses.
    B.   **Numbers of microorganisms initially present**
    C.   **Environmental conditions** affect death rate of microorganisms
        1.   pH
        2.   Presence of fats and other organic compounds
    D.   **Potential risk of infection**
5.3.   Using heat to destroy microorganisms and viruses
    A.   **Moist heat**—Moist heat, such as boiling water and pressurized steam, destroys microorganisms by causing the irreversible coagulation of their proteins.
    B.   **Dry heat**—Dry heat, such as in direct flaming and ovens destroy microorganisms by oxidizing cells to ashes or irreversibly denaturing their proteins.

5.4   Using chemicals to destroy microorganisms and viruses
    A.   **Germicidal chemicals** can be used to disinfect and, in some cases, sterilize, but they are less reliable than heat. Most chemical germicides react irreversibly with vital enzymes and other proteins, the cytoplasmic membrane, or viral envelopes.
    B.   **Potency** of germicidal chemical formulations
        1.   Sterilants
        2.   High-level disinfectants
        3.   Intermediate-level disinfectants
        4.   Low-level disinfectants
    C.   **Selection** factors for the appropriate germicidal chemical
        1.   Toxicity
        2.   Residue
        3.   Activity in the presence of organic matter
        4.   Compatibility with the material being treated
        5.   Cost and availability
        6.   Storage and stability
        7.   Ease of disposal
    D.   **Classes** of germicidal chemicals
        1.   **Ethyl or isopropyl alcohol** (60-80% solution) in water rapidly kills vegetative bacteria and fungi by coagulating enzymes and other essential proteins, and by damaging lipid membranes.
        2.   **Gluteraldehyde** and **formaldehyde** destroy microorganisms and viruses by inactivating proteins and nucleic acids. A 20% solution of alkaline gluteraldehyde is one of the most widely used chemical sterilants.
        3.   **Chlorhexidine** is a **biguanide** extensively used in antiseptic products.
        4.   **Ethylene oxide** is a gaseous sterilizing agent that penetrates well and destroys microorganisms and viruses by reacting with proteins.
        5.   **Sodium hypochlorite** (liquid bleach) is one of the least expensive and most readily available forms of chlorine. Chlorine dioxide is used as a sterilant and disinfectant. Iodophores are iodine-releasing compounds used as antiseptics.
        6.   **Metals** interfere with protein function. Silver-containing compounds are used to prevent wound infections.
        7.   **Ozone** is used as an alternative to chlorine disinfection of drinking water and wastewater.
        8.   **Peroxide** and **peracetic acid** are both strong oxidizing agents that can be used alone or in combination as sterilants.
        9.   **Phenolic compounds** destroy cytoplasmic membranes and denature proteins.
       10.   **Quaternary ammonium compounds** are cationic detergents that are non-toxic enough to be used to disinfect food preparation surfaces.

5.5   Removal of microorganisms by filtration
    A.   Filtration of fluids
        1.   **Depth filters** have complex, tortuous passages that retain microorganisms while letting the suspending fluid pass through the small holes.
        2.   **Membrane filters** are produced with graded pore sizes extending below the dimensions of the smallest known viruses.
    B.   Filtration of air
        1.   **High efficiency particulate air (HEPA) filters** remove nearly all microorganisms.
        2.   HEPA filters are used in specialized hospital rooms to protect patients, biological safety cabinets and laminar flow hoods.

5.6   Using **radiation** to destroy microorganisms and viruses
   A.   **Gamma irradiation** cause biological damage by producing superoxide and hydroxyl free radicals. Irradiation can be used to:
      1.   Sterilize heat-sensitive materials
      2.   Decrease the numbers of microorganisms in foods.
   B.   **Ultraviolet light** is used to disinfect surfaces by damaging nucleic acids by causing the formation of covalent bonds between adjacent thymine molecules in DNA, creating **thymine dimers**.
   C.   **Microwaves** do not affect microorganisms directly but they can kill microorganisms by the heat they generate in a product.

5.7   **Preservation** of perishable products by techniques that slow or halt the growth of microorganisms to delay spoilage.
   A.   **Chemical preservatives**
      1.   Organic acids such as benzoic, sorbic and propionic acids
      2.   Nitrate and nitrite
   B.   **Low temperature storage**
      1.   Low temperatures above freezing inhibit microbial growth.
      2.   Freezing essentially stops all microbial growth.
   C.   **Reducing the available water** by addition of sugars and salts
   D.   **Lyophilization** is used for preserving food

Terms You Should Know

Antiseptic	Disinfection	Pasteurization
Aseptic technique	Fungicide	Preservation
Autoclave	Germicide	Sanitize
Bactericidal	HEPA filter	Sterile
Bacteriostatic	Laminar flow hoods	Sterilization
Decimal reduction time	Lyophilization	Viricide
Decontamination	Normal flora	
Disinfectant	Nosocomial infections	

Learning Activities

1.   To what are each of these microbes resistant?

Endospores of *Bacillus* and *Clostridium*	
Mycobacterium species	
Pseudomonas species	
Naked viruses	

2. List four considerations for selecting an antimicrobial procedure.

3. If the D value of a microorganism were 3 minutes, how long would it take to reduce a population of 1000 microorganisms to only one surviving organism?

4. Describe the differences among the three kinds of pasteurization methods.

Method	
Low-temperature-long-time method (62°C)	
High-temperature-short-time method (72°C)	
Ultra-high-temperature method (140-150°C)	

5. Indicate which of the following methods of microbial control will kill vegetative cells and which will kill endospores.

Method of microbial control	
Autoclave	
Pasteurization	
Boiling	
Dry heat	
Freezing	

6. Which of the following methods of sterilization can be used to sterilize the following?

Method of sterilization	Heat-labile solution	Liquids	Glassware
Dry heat			
Autoclave			
Membrane filtration			
Pasteurization			

7. Give the advantages of the following methods.

Sterilization Method	
Autoclave	
Pasteurization	
Hot-air oven	
Direct flaming	
Filtration	
Irradiation with gamma ray	
Ultraviolet light	

8. Describe seven considerations in selection of an appropriate germicidal chemical.

9. Describe the use of each of the following antimicrobial chemicals.

Chemical	Use
Alcohols	
Aldehydes	
Biguanides	
Ethylene oxide gas	
Halogens	
Metals	
Ozone	
Peroxygens	
Phenolic compounds	
Quaternary ammonium compounds	

10. Explain the advantages and disadvantages of each of the following as a surgical scrub.

Antiseptic/disinfectant	Explanation
Dial soap	
Chlorine bleach	
Phenol	
Chlorhexidine	
Glutaraldehyde	

11. Describe the use of each of the following antimicrobial methods.

Method	
Membrane filters	
HEPA filters	
Gamma radiation	
Ultraviolet radiation	

12. Which of the following antimicrobial agents can be used to retard the growth in microbes in foods?

Antimicrobial agent	Retards growth in foods
Sodium nitrate	
Glutaraldehyde	
Alcohols	
Sorbic acid *	
Sodium benzoate	
Calcium propionate	

*An organic acid

Self Test

1. Which of the following would be the most efficient way to kill endospores?

   a. autoclave at 121°C for 15 minutes
   b. boiling
   c. hot air oven at 180°C for 3 hours
   d. pasteurization
   e. none of the above

2. A substance used on human tissues to kill microbes is called a(n)

   a. antibiotic.
   b. antiseptic.
   c. disinfectant.
   d. sterilant.
   e. all of the above

3. Thymine dimers can be formed in cellular DNA by which of the following?

   a. nitrous acid
   b. ultraviolet light
   c. gamma rays
   d. base analogs
   e. X-rays

4. Which concentration of ethyl alcohol is the most effective as a disinfectant?

   a. 100%
   b. 70%
   c. 50%
   d. 30%
   e. They are all equal in effectiveness

5. Which of the following statements *best* describes how microbes die in a population when they are exposed to an antimicrobial substance?

   a. Death of microbes depends on the species.
   b. The microbes all die at one time.
   c. Death of microbes depends on the antimicrobial agent.
   d. All of the microbes are not ever killed.
   e. The microbes in a population die in an exponential manner.

6. Which of the following statements best describes the effect of temperature on the effectiveness of a disinfectant?

   a. Warmer temperatures enhance effectiveness.
   b. Colder temperatures enhance effectiveness.
   c. Warmer temperatures reduce effectiveness.
   d. The effectiveness of a disinfectant mainly depends upon the type of microbe.
   e. Temperature has nothing to do with the effectiveness of a disinfectant.

7. A tincture is a solution of a substance dissolved in

   a. water.
   b. alcohol.
   c. iodine.
   d. gas.
   e. none of the above

8. Which of the following is relatively ineffective as a disinfectant?

   a. phenol
   b. alcohol
   c. iodine
   d. soap
   e. hydrogen peroxide

9. Which of the following can be used to sterilize heat or moisture sensitive items?

   a. alcohol
   b. chlorine
   c. soap
   d. phenol
   e. ethylene oxide

10. Which of the following is the most effective in killing bacteria?

    a. autoclaving
    b. boiling
    c. extended periods in a hot air oven
    d. freezing
    e. repeated freezing and thawing

Thought Questions

1. Why are some microorganisms easier to kill than others?

2. If in the sterilization process half of a population of bacteria were killed in the first minute, what proportion of the original population would be killed in the second minute?

Answers to Self Test Questions

1-a, 2-b, 3-b, 4-b, 5-e, 6-a, 7-b, 8-d, 9-e, 10-a

Chapter 6   Metabolism: Fueling Cell Growth

Overview

Harvesting energy is essential for the life and functioning of cells. In this chapter the principles of metabolism are presented. Enzymes are described and their role in metabolism is discussed. Specific metabolic pathways are described with an emphasis on the central metabolic pathways. Respiration and fermentation are discussed. Photosynthesis and carbon fixation also are presented. Anabolic pathways are summarized.

Learning Objectives

After studying the material in this chapter, you should be able to:

1. Differentiate between the two components of metabolism—catabolism and anabolism.
2. Distinguish between potential and kinetic energy.
3. Distinguish between exergonic and endergonic reactions.
4. Define oxidation-reduction reactions.
5. Describe the roles in metabolic pathways of the following:
    - Enzymes
    - ATP
    - Energy source
    - Electron carriers
    - Precursor metabolites
    - Activation energy
6. Define enzyme and describe how enzymes function.
7. Describe how enzymes are controlled by inhibition (allosteric regulation).
8. Describe the role of cofactors and coenzymes in enzyme function.
9. List and explain the factors influencing enzymatic activity.
10. Distinguish between competitive and noncompetitive inhibition.
11. List the central metabolic pathways.
12. Define:
    - Amphibolic pathways
    - Entner-Doudoroff pathway
    - Pentose phosphate pathway
13. Define each of the following and describe what occurs in each:
    - Glycolysis
    - Transition step (oxidative decarboxylation of pyruvate)
    - Tricarboxylic acid cycle
14. Describe precisely where ATP is made.
15. Explain why 38 ATPs can be theoretically produced from the oxidation of one molecule of glucose in the prokaryotic cell whereas only 36 ATPs may be produced from the same process in the eukaryotic cell.
16. Oxidative phosphorylation
17. Define respiration and differentiate between aerobic and anaerobic respiration.

18. Define proton motive force and describe how it functions.
19. List the carriers involved in the electron transport chain.
20. Describe the role of electron carriers.
21. List the protein complexes in mitochondria and indicate which ones function as proton pumps.
22. Explain how ATP synthase harvests the proton motive force to synthesize ATP.
23. Define fermentation and list some of its important products.
24. Describe how the following types of organisms harvest energy:
    - Chemolithotrophs
    - Photosynthetic organisms
25. Define photosynthesis.
26. Differentiate among the following:
    - Cyclic photophosphorylation
    - Non-cyclic photophosphorylation
    - Calvin cycle
27. Explain how lipid catabolism and protein catabolism can be integrated with carbohydrate metabolism.
28. Describe how the following compounds can synthesized through anabolic reactions:
    - subunits of lipids
    - amino acids
    - nucleotides

Key Concepts

1. Metabolism is all the chemical reactions that occur in a cell.
2. Cells break down (catabolize) nutrients in a controlled manner in order to (1) obtain energy that can be used for cellular activity and (2) obtain the biochemical units to build new molecules for use in the cell.
3. Cells build up (anabolism) new molecules in order to (1) store energy and (2) to make new molecules used in cellular metabolic pathways or in cell structure.
4. Nutrients are broken down by the central metabolic pathways, which are: (1) glycolysis, (2) the pentose-phosphate pathway and (3) the TCA (Krebs) cycle.
5. The electron transport chain is a series of electron carrier molecules that sequentially pass electrons from one to another in order to produce a proton motive force used to fuel the synthesis of ATP which is the immediate source of energy used in cell activities.
6. In aerobic respiration the terminal electron acceptor is oxygen, while in anaerobic respiration the terminal electron acceptor is an inorganic compound other than oxygen. Aerobic respiration results in a greater ATP yield than does anaerobic respiration.
7. Fermentation is a process in which the terminal electron acceptor is an organic molecule. The TCA cycle and electron transport chain are not used.
8. Enzymes (biological catalysts) are used to catalyze chemical reactions in the cell.
9. Bacteria vary in the kinds and numbers of enzymes that they use to synthesize various products. Specific enzymes and products can be used in the identification of bacterial species.
10. Bacteria vary in the sources they use for energy and carbon.

Summary Outline

6.1   Principles of metabolism
- A.   **Catabolism** encompasses those processes that transform and release energy.
- B.   **Anabolism** includes the processes that utilize energy to synthesize and assemble the building blocks of a cell.
- C.   Harvesting energy
  - 1.   **Energy** is the ability to do work.
  - 2.   The first law of thermodynamics states that the energy in a system can never be created or destroyed.
  - 3.   The second law of thermodynamics states that **entropy** always increases.
  - 4.   **Photosynthetic organisms** harvest the energy of sunlight, using it to power the synthesis of organic compounds.
  - 5.   **Chemoorganotrophs** transform energy by organic compounds.
  - 6.   Free energy is the amount of energy that can be gained by breaking the bonds of a chemical.
    - a)   **Exergonic** reactions release energy.
    - b)   **Endergonic** reactions utilize energy.
- D.   Components of metabolic pathways
  - 1.   A **specific enzyme** facilitates each step of a metabolic pathway by **lowering the activation energy** of a reaction that converts a substrate into a product.
  - 2.   **ATP** is the energy currency of the cell.
    - a)   **Substrate level phosphorylation** uses the chemical energy released in an exergonic reaction to add P_i to ADP.
    - b)   **Oxidative phosphorylation** harvests the energy of proton motive force to do the same thing.
  - 3.   The energy source is oxidized to release its energy; this oxidation-reduction reaction **reduces an electron carrier**.
  - 4.   **NAD^+, $NADP^+$,** and **FAD** are electron carriers. Their reduced form functions as reducing power. NADH and $FADH_2$ are used to provide electrons for the generation of proton motive force. NADPH is used in biosynthesis.
  - 5.   **Precursor metabolites** are building blocks that can be used to make the subunits of macromolecules, but they can also be oxidized to release energy.
- E.   Scheme of metabolism
  - 1.   The central metabolic pathways are:
    - a)   **Glycolysis**
    - b)   The **pentose phosphate pathway**
    - c)   **The tricarboxylic acid cycle (TCA cycle)**
  - 2.   **Glycolysis** oxidizes glucose to pyruvate, producing ATP, reducing power and precursor metabolites.
  - 3.   The **pentose phosphate pathway** also oxidizes glucose to pyruvate, but its primary role is the production of precursor metabolites and reducing power essential for biosynthesis.
  - 4.   The **transition step** forms acetyl CoA, which then enters the tricarboxylic acid cycle (TCA) cycle.
  - 5.   **Respiration** uses the reducing power accumulated in the central metabolic pathways to generate ATP by oxidative phosphorylation.
    - a)   **Aerobic respiration** uses O_2 as a terminal electron acceptor.
    - b)   **Anaerobic respiration** uses an inorganic molecule other than O_2 as a terminal electron acceptor.

     6.   **Fermentation** uses pyruvate or a derivative as a terminal electron acceptor rather than oxidizing it further in the TCA cycle; this recycles the reduced electron carrier NADH.

6.2   Enzymes
- A.   **Enzymes** function as **biological** catalysts, which are not permanently changed during a reaction.
- B.   The enzyme **substrate** binds to the active site or catalytic site to form a temporary intermediate called an enzyme-substrate complex.
- C.   **Cofactors** and **coenzymes** act in conjunction with enzymes.
- D.   **Environmental factors** that influence enzyme activity include
  1. Temperature
  2. pH
  3. salt concentration
- E.   **Allosteric regulation** uses an effector that binds to the allosteric site of the enzyme that in turn alters the relative affinity of the enzyme for its substrate.
- F.   **Enzyme inhibition**
  1. **Competitive inhibition** occurs when the inhibitor competes with the normal substrate for the active binding site.
  2. **Non-competitive inhibition** occurs when the inhibitor and the substrate act as different sites on the enzyme.

6.3   Central metabolic pathways
- A.   **Glycolysis** is a nine-step pathway that converts one molecule of glucose into two molecules of pyruvate; the theoretical net yield is two ATP, two NADH + H^+ and six different precursor metabolites.
- B.   **Pentose phosphate pathway** can generate some ATP, but its greatest significance is that it forms NADPH and two different precursor metabolites.
- C.   **Transition step** results in the decarboxylation and oxidization of pyruvate, and joins the resulting acetyl group to coenzyme A forming acetyl-Co A. This produces NADH + H^+ and one precursor metabolite.
- E.   **Tricarboxylic acid cycle** completes the oxidation of glucose; the theoretical yield is 6 NADH + $6H^+$, $2FADH_2$, 2ATP and three different precursor metabolites.

6.4   Respiration
- A.   The **reducing power** accumulated in **glycolysis, the transition step,** and the **TCA cycle** is used to drive the **synthesis of ATP.**
- B.   **Electron transport chain**
  1. The **electron transport chain** sequentially passes electrons, and, as a result, ejects protons. Most of the carriers are grouped into large complexes that function a proton pumps that generate the **chemiosmotic gradient** called the **proton motive force**.
  2. The mitochondrial electron transport chain has three different complexes (complexes I, II, and IV) that function as proton pumps.
  3. Prokaryotes vary with respect to the types and arrangements of their electron transport components.
  4. Some prokaryotes can use inorganic molecules other than O_2 as a terminal electron acceptor (**anaerobic respiration**). This harvests less energy than aerobic respiration.
- C.   **ATP synthetase**
  1. ATP synthetase harvests the energy released by the electron transport chain as it allows protons to move back across the membrane, driving the synthesis of ATP.
  2. The theoretical maximum yield of ATP of aerobic respiration is 38 ATP.

6.5   **Fermentation** is used by organisms that cannot respire, either because a suitable inorganic terminal electron acceptor is not available or because they lack an electron transport chain.

6.6.   Catabolism of organic compounds other than glucose
- A.   Hydrolytic enzymes break down macromolecules into their respective subunits.

B.   **Polysaccharides and disaccharides**
1.   Amylases digest starch, releasing glucose subunits.
2.   Cellulases degrade cellulose.
3.   The sugar subunits can enter glycolysis to be oxidized to pyruvate.
C.   **Lipids**
1.   Fats are hydrolyzed by lipase, releasing glycerol and fatty acids.
2.   Glycerol is converted to the precursor metabolite glyceraldehyde 3-phosphate; fatty acids are degraded by beta-oxidation, generating reducing power and the precursor metabolite acetyl-CoA.
D.   **Proteins**
1.   Proteins are hydrolyzed by proteases.
2.   Deamination removes the amino group; the remaining carbon skeleton is then converted into the appropriate precursor molecule.

6.7   **Chemolithotrophs** are autotrophs; they do not require an external source of organic carbon because they can fix carbon dioxide.

6.8   **Photosynthesis** captures the energy of sunlight and uses it to drive the synthesis of ATP.
A.   The role of photosynthetic pigments
1.   **Chlorophylls** are the primary pigments used to harvest solar energy.
2.   **Carotenoids** are accessory pigments that absorb wavelengths of light not absorbed by the chlorophylls and then transfer that energy to the chlorophylls.
B.   **Photophosphorylation**—light energy excites an electron, which is passed along an electron transport chain, generating proton motive force.
C.   Electron source
1.   **Oxygenic phototrophs** extract electrons from water.
2.   **Anoxygenic phototrophs** extract electrons from reduced compounds other than water.

6.9   Carbon fixation—**Calvin cycle** is used to incorporate CO_2 into organic carbon.

6.10   **Anabolic pathways**—synthesizing subunits from precursor molecules
A.   Lipid synthesis
B.   Amino acid synthesis
B.   Nucleotide synthesis

Terms You Should Know

Activation energy	Electron carrier	Oxidation-reduction reactions
Active site	Electron donor	Oxidative phosphorylation
Aerobic respiration	Electron transport chain	Pentose phosphate pathway
Allosteric site	End product	Photophosphorylation
Amphibolic pathway	Endergonic	Photosynthesis
Anabolism	Energy	Potential energy
Anaerobic respiration	Energy source	Precursor metabolites
ATP	Entner-Doudoroff pathway	Products
ATP synthase	Enzyme	Proton motive force
Biosynthesis	Enzyme-substrate complex	Proton pump
Calvin cycle	Exergonic	Reactants
Carbon fixation	Feedback inhibition	Respiration
Catabolism	Fermentation	Respiration
Central metabolic pathways	Free energy	Substrate
Chemiosmotic theory	Glycolysis	Substrate level phosphorylation
Chemolithotrophs	Intermediates	Terminal electron acceptor
Chemoorganotrophs	Kinetic energy	Transition step
Coenzymes	Metabolic pathway	Tricarboxylic acid cycle (TCA)
Cofactors	Metabolism	
Competitive inhibition	Non-competitive inhibition	

Learning Activities

1. Define:

Metabolism	
Anabolism	
Catabolism	
Metabolic pathway	

2. Describe what occurs in oxidation-reduction reactions and explain how they are coupled.

Reaction	What occurs
Oxidation	
Reduction	

3. Identify the function of the following compounds.

Compound	Function
ATP (adenosine triphosphate)	
ADP (adenosine diphosphate)	
NAD^+ (nicotinamide adenine dinucleotide)	
FAD (flavin adenine dinucleotide)	
$NADP^+$ (nicotinamide adenine dinucleotide phosphate)	
NADH (reduced form of NAD^+)	
$FADH_2$ (reduced form of FAD)	
NADPH (reduced form of $NADP^+$)	
Pyruvate	
Chorophylls	
Carotenoids	

4. Define each of the following.

Component	Definition
Enzyme	
Coenzyme	
Cofactor	
Substrate	
Enzyme-substrate complex	
Product	
Activation energy	

5. Describe how an enzyme (biological catalyst) works.

6. Describe using a diagram allosteric regulation of enzymes.

7. Define enzyme inhibition.

_____

_____

_____

8. Differentiate between competitive and noncompetitive inhibition.

Competitive inhibition	
Noncompetitive inhibition	

9. ATP can be made in the three ways listed below. Match them with specific examples: (You may use choices more than once, only once or not at all.)

	1. NAD + 2e⁻+ 2H⁺→ NADH + H⁺	A. Substrate-level phosphorylation
	2. NADH + H⁺ → NAD +2e⁻ + 2H⁺	B. Oxidative phosphorylation
	3. Isocitric acid to α-ketoglutaric acid *	C. Photophosphorylation
	4. A phosphate group is transferred from an organic molecule to ADP	
	5. ATP and NADPH are made in a plant cell	

*Look at the TCA cycle

10.   Match the type of phosphorylation with the true statements.

	Statement	Type of Phosphorylation
	1. A high-energy phosphate group is transferred from an intermediate to ADP	A. Substrate-level phosphorylation
	2. Occurs in glycolysis	B. Oxidative phosphorylation
	3. No final electron acceptor is required	C. Photophosphorylation
	4. Energy is released as carrier molecules are oxidized to transform energy to generate ATP	D. None of these
	5. Transfer of a high-energy phosphate group to ADP	
	6. Requires CO_2	
	7. Requires light	
	8. Protons are moved across a membrane and provide energy to phosphorylate ADP	

11. Describe the purpose of the pentose phosphate pathway.

_____

_____

_____

12. List the three central metabolic pathways.

13. Complete the table for the following processes:

Process	Starting compound	End products	Energy carriers produced such as NADH, FADH$_2$ or NADPH
Glycolysis			
Transition step			
TCA Cycle			
Electron transport chain			
Fermentation			
Pentose phosphate cycle			
Anaerobic respiration			

14. For the following processes give the amounts of ATP directly and ultimately (via the electron transport chain) that can be theoretically produced.

Process	Molecules of ATP produced directly	Molecules of ATP produced ultimately
Glycolysis		
Transition step		
TCA cycle		
Electron transport chain		
Calvin cycle		
Pentose phosphate cycle		
Anaerobic respiration		

15. How many molecules of ATP are ultimately formed (via the electron transport chain and directly) by the following reactions or series of reactions:

Reactions	ATP ultimately formed
Glucose → Glucose 6-phosphate	
Phosphoenolpyruvaate → Pyruvate	
Glucose → Pyruvate	
Acetyl → CO_2 + H_2O	
Succinate → Fumarate	
Glucose → CO_2 + H_2O (Prokaryotic cell)	
Glucose → CO_2 + H_2O (Eukaryotic cell)	

16. Describe the role of the electron motive force in harvesting energy.

17. The two types of cell respiration and fermentation can be differentiated by the type of terminal electron acceptors used. Name the kind of compound that constitutes the terminal electron acceptor for the following:

Process	Terminal Electron Acceptor
Aerobic respiration	
Anaerobic respiration	
Fermentation	

18. How does a strictly fermentative bacterium transform energy?

_____

_____

_____

19. List the type of microorganisms that might be identified by the following fermentation end products.

End product	Microorganism
Lactic acid	
Ethanol	
Butryic acid	
Butanediol	
Mixed acids	

20. The following compounds have varying amounts of potential energy. Rank them from the most potential energy (1) to the least (7).

Compound	Amount of Potential Energy
Glucose	
Pyruvic acid	
Starch	
Acetyl	
ADP	
ATP	
CO_2	

21. Define photosynthesis:

_____

_____

_____

22. Distinguish between cyclic photophosphorlyation and non-cyclic photophosphorlyation.

_____

_____

_____

23. Give the subunits of the following compounds:

Compounds	Component molecules
Polysaccharides	
Lipids	
Proteins	
Nucleic acids	

24. For each of the four nutritional types of microorganisms give the energy and carbon sources.

Nutritional Types	Energy source	Carbon source
Chemolithotrophs		
Chemoorganootrophs		
Photosynthetic organisms		

25. List the precursor molecules for the following subunits.

Subunit	Precursor molecules
Fatty acids	
Glycerol	
Glutamate	
Aromatic amino acids	
Nucleotides	

26.   On the scheme of metabolism drawing label the following:

Glycolysis, TCA cycle, respiration, fermentation, transition step, pentose phosphate pathway

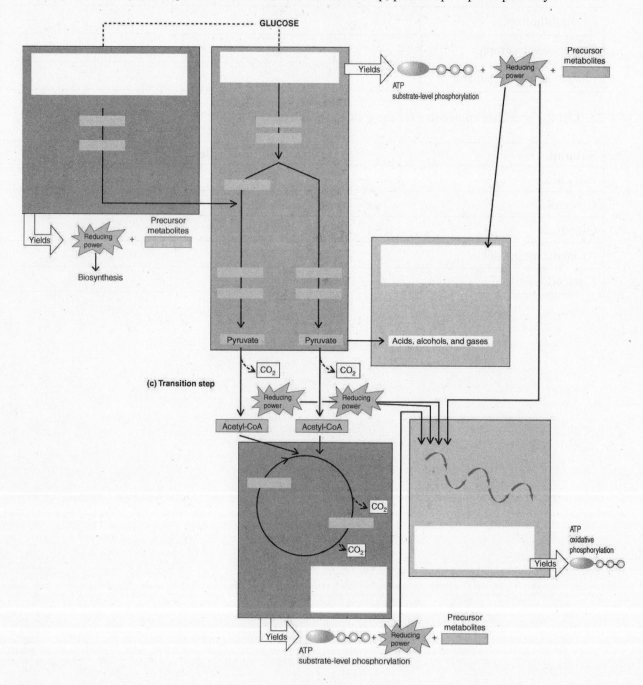

27. Label the events of glycolysis.

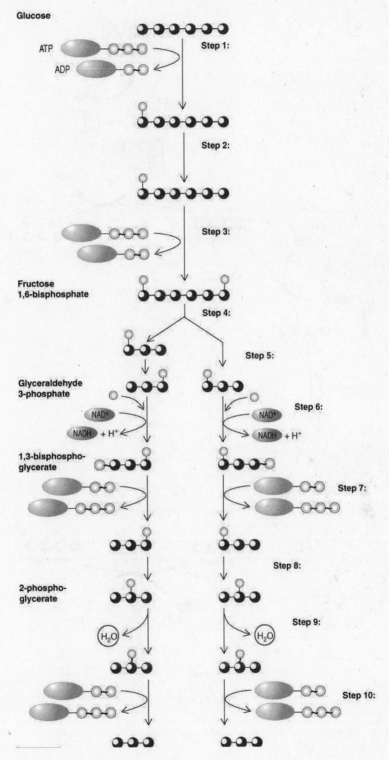

Glucose

ATP

ADP

Step 1:

Step 2:

Step 3:

Fructose
1,6-bisphosphate

Step 4:

Step 5:

Glyceraldehyde
3-phosphate

NAD⁺

NAD⁺

Step 6:

NADH + H⁺

NADH + H⁺

1,3-bisphospho-
glycerate

Step 7:

2-phospho-
glycerate

Step 8:

H₂O

H₂O

Step 9:

Step 10:

28. Label the events of the TCA cycle.

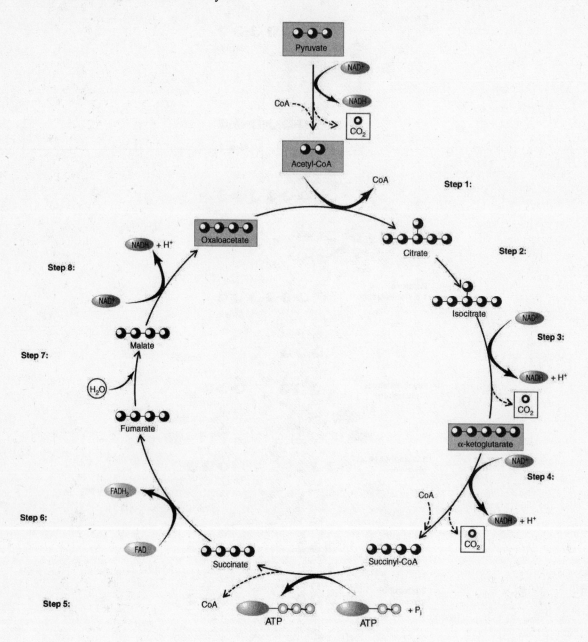

29. On the drawing illustrating the mechanism of enzyme action, label the following:

Substrate, enzyme, enzyme-substrate complex, products, active site

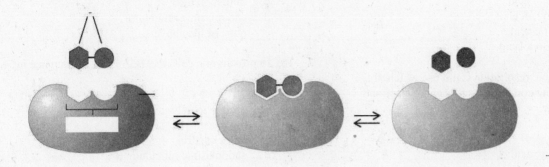

Self Test

1. In aerobic respiration the terminal electron acceptor(s) is/are

   a. nitrates and sulfates.
   b. oxygen.
   c. an organic compound.
   d. pyruvic acid.
   e. a cytochrome.

2. Reduction is the

   a. addition of protons.
   b. addition of electrons or hydrogen atoms.
   c. addition of oxygen.
   d. removal of protons.
   e. removal of electrons or hydrogen atoms.

3. The amount of ATP synthesized during aerobic respiration is

   a. much less than that produced in fermentation.
   b. greater than that produced in fermentation.
   c. about the same as that produced in fermentation.

4. How many molecules of ATP may be produced from the complete oxidation of one molecule of glucose in bacteria?

   a. 2
   b. 4
   c. 36
   d. 38
   e. 40

5. Which of the following is not an enzyme?

   a. lactase
   b. lipase
   c. dehydrogenase
   d. coenzyme A
   e. hexokinase

6. ATP is produced in glycolysis by what method?

   a. substrate phosphorylation
   b. oxidative phosphorylation
   c. photophosphorylation
   d. chemiosomosis

7. The phosphorylation process involved in the production of ATP from ADP and inorganic phosphate with the transfer of electrons and energy is called:

   a. substrate phosphorylation
   b. oxidative phosphorylation
   c. photophosphorylation

8. Molecules on which enzymes act are called

   a. cofactors.
   b. coenzymes.
   c. substrates.
   d. inhibitors.
   e. activated molecules.

9. Products of fermentation may include all of the following except

    a. carbon dioxide.
    b. acetic acid.
    c. ethyl alcohol.
    d. lactic acid.
    e. pyruvic acid.

10. Anaerobic respiration could use which of the following compounds as terminal electron acceptor.

    a. oxygen
    b. lactic acid
    c. pyruvate
    d. cytochrome
    e. nitrate

11. Oxidation involves the

    a. addition of protons.
    b. addition of electrons or hydrogen atoms.
    c. removal of electrons or hydrogen atoms.
    d. removal or protons.
    e. addition of oxygen atoms.

12. In the fermentation process, glucose is oxidized to

    a. carbon dioxide and water.
    b. pyruvate.
    c. a variety of products depending on the microorganism.
    d. none of the above because fermentation is not an oxidative process.

13. Enzymes function by

    a. reducing the required activation energy of the reaction.
    b. increasing the required activation energy of the reaction.
    c. reducing random molecular motion.
    d. increasing random molecular motion.
    e. increasing the temperature of the reaction.

14. In prokaryotic cell, glycolysis takes place in the

    a. plasma membrane.
    b. cytoplasm.
    c. mitochondria.
    d. endoplasmic reticulum.
    e. none of the above

15. In prokaryotic cells the TCA cycle takes place in the

    a. plasma membrane
    b. cytoplasm
    c. mitochondria
    d. endoplasmic reticulum
    e. none of the above

16. In eukaryotic cells, glycolysis takes place in the

    a. plasma membrane.
    b. cytoplasm.
    c. mitochondria.
    d. endoplasmic reticulum.
    e. none of the above

17. In eukaryotic cells the TCA cycle takes place in the

    a. plasma membrane
    b. cytoplasm
    c. mitochondria
    d. endoplasmic reticulum
    e. none of the above

Thought Questions

1. Describe the general purposes of metabolic reactions, both anabolic and catabolic.

2. Distinguish among substrate phosphorylation, oxidative phosphorylation and photophosphorylation.

3. Describe how proton motive force is generated.

4. Explain why glycolysis, tricarboxylic acid cycle, and the electron transport chain called the central metabolic pathways.

Answers to Self Test Questions

1-b, 2-b, 3-b, 4-d, 5-d, 6-a, 7-b, 8-c, 9-e, 10-a, 11-c, 12-a, 13-a, 14-b, 15-a, 16-b, 17-c

Chapter 7   The Blueprint of Life, from DNA to Protein

Overview

For a cell to carry on its normal functions the information encoded in the DNA must be made available to the ribosomes for the production of proteins. The nature of DNA and RNA is discussed in this chapter as well as the processes of (1) replication or copying of the DNA, (2) transcription or copying the information in the DNA into RNA and (3) translation or using the encoded information in RNA to make a specific protein. The differences between eukaryotic and prokaryotic gene expression are discussed as well as the principles of regulation of gene expression.

Learning Objectives

After studying the material in this chapter, you should be able to:

1. Define:
   - Leading strand
   - Lagging strand
   - Minus (-) strand of DNA
   - Plus (+) strand of DNA
   - DNA polymerase
2. Describe DNA replication. Include initiation, a description of the replication fork and the enzymes involved.
3. Define:
   - RNA polymerase
   - mRNA
   - rRNA
   - tRNA
4. Describe the transcription of DNA to RNA.
5. Describe the translation of RNA into protein.
6. Define:
   - Genetic code
   - Codon
   - Stop codon
   - Anticodon
7. Given the codon (mRNA) for specific amino acids determine:
   - The order of bases in the minus strand of DNA.
   - The order of bases in the plus strand of DNA.
   - The anticodon for each specific amino acid.
8. Explain what is meant by inducible and repressible enzymes.
9. Describe the operon model and explain how it controls the production of proteins.

10. Identify and give the role of the
    - Operator
    - Promoter
    - Ribosome binding site
11. Define and describe the significance of
    - Quorum sensing
    - Genomics
12. Identify the function of the following compounds.
    - Deoxyribonucleic acid (DNA)
    - Ribonucleic acid (RNA)
    - Messenger RNA (mRNA)
    - Ribosomal RNA (rRNA)
    - Transfer RNA (tRNA)
    - DNA polymerases
    - DNA ligase
    - DNA gyrase
    - RNA polymerase

Key Concepts

1. DNA is a double-stranded helical molecule with antiparallel strands (they are oriented in opposite directions).
2. RNA is a single-stranded molecule that is transcribed or copied from one of the strands of DNA.
3. Replication is the process of duplicating or copying the double-stranded DNA molecule.
4. Transcription is the process of copying the information encoded in the DNA into mRNA.
5. Translation is the process of utilizing the information encoded in the mRNA to construct a specific protein.
6. Gene expression, which involves both the transcription of DNA and the translation of RNA into protein, is controlled by specific mechanisms.
7. Expression of genes is moderated in response to changes in environmental conditions.
8. Genomics is science that determines the sequences of nucleotides in the genes of an organism. Sequencing methodologies have become more rapid, but analyzing the data and extracting the pertinent information is still difficult.

Summary Outline

7.1   Overview
    A.   Definitions
        1. A **genome** is the complete of genetic information for a cell.
        2. **Replication** is the process of duplicating double-stranded DNA.
        3. **Transcription** is the process of copying the information encoded on DNA into RNA.
        4. **Translation** is the process of interpreting the information carried by messenger RNA in order to synthesize the encoded protein.
    B.   Characteristics of **DNA**
        1. A single strand of **DNA has a 5 prime end and a 3 prime end**.
        2. The **two strands of DNA** in the **double helix are antiparallel**.
        3. The separating of double-stranded DNA is called **denaturing** or **melting**.

C.   Characteristics of **RNA**
1.   A **single-stranded RNA** fragment is transcribed from one of the two strands of DNA.
2.   There are three different functional groups of RNA molecules:
      a)   **Messenger RNA (mRNA)**
      b)   **Ribosomal RNA (rRNA)**
      c)   **Transfer RNA (tRNA)**

D.   **Regulating** the **expression of genes**
1.   Protein synthesis is generally controlled by regulating the synthesis of mRNA molecules.
2.   mRNA is short-lived because RNases degrade it within minutes.

7.2   **DNA replication**
A.   DNA replication is generally **bi-directional**.
B.   Replication of double-stranded DNA is **semiconservative**.
C.   The DNA chain always elongates in the **5′ to 3′ direction**.
D.   **Base pairing** rules determine the **specific nucleotides** that are added.
E.   DNA replication begins at the origin of replication.
F.   **DNA polymerase** synthesizes DNA in the 5′ to 3′ direction, using one strand as a **template** to generate the **complementary strand**.
G.   The **bi-directional progression of replication** around a circular DNA molecule creates two **replication forks**.
H.   Enzymes involved in DNA replication include:
1.   **DNA polymerase**
2.   **Primase**
3.   **DNA ligase**
4.   **DNA gyrase**

7.3   Gene expression
A.   **Transcription**
1.   **RNA polymerase** catalyzes transcription, producing a single-stranded RNA molecule that is complementary and antiparallel to the DNA template.
2.   In prokaryotes, an mRNA molecule can be **monocistronic** or **polycistronic**.
3.   Transcription begins when RNA polymerase recognizes and binds to a sequence of nucleotides on the DNA called a **promoter**; it is the **sigma subunit** of the enzyme that recognizes the promoter sequence.
4.   **RNA is synthesized in the 5′ to 3′ direction**.
5.   A **transcription terminator** causes RNA polymerase to fall off the DNA template and to release the newly synthesized RNA.

B.   **Translation**
1.   The information encoded on mRNA is deciphered using the **genetic code**.
2.   **Ribosomes** are the **sites at which translation** occurs.
3.   **tRNAs carry specific amino acids**.
4.   **Initiation of translation** begins when **the ribosome binds to the ribosome-binding site** on the mRNA molecule; this occurs even while the mRNA is still being synthesized. **Translation starts** at the **first AUG downstream** of that site.
5.   The ribosome moves along the mRNA in the **5′ to 3′ direction** so that **one codon is translated at a time**. Translation terminates when the ribosome reaches a **stop codon**.
6.   Proteins are often modified after they are synthesized.

7.4   **Differences** between **eukaryotic** and **prokaryotic gene expression**
A.   Eukaryotic mRNA is processed; **a cap and a poly A tail are added**.
B.   Eukaryotic genes often contain **introns** that are **removed** from precursor mRNA by a process called **splicing**.

C.   In eukaryotic cells, the **mRNA must be transported** out of the nucleus before it can be translated in the cytoplasm. Eukaryotic mRNA is **typically monocistronic**.

7.5   **Genomics** DNA sequence is analyzed and compared to other known sequences by searching a computerized database.

7.6   **Regulating gene expression**
    A.   Principles
        1.   Genes encoding **constitutive enzymes** are always active.
        2.   Genes encoding enzymes that can be **induced** are turned on only by certain conditions; those that can be **repressed** are turned off by certain conditions.
    B.   Mechanisms to **control** transcription
        1.   Global control is the simultaneous regulation of numerous genes unrelated in function.
        2.   Many genes have a **regulatory region** near their promoter to which a specific protein can bind, controlling transcription.
        3.   An **operon** is a set of adjacent genes coordinately controlled by a regulatory protein and transcribed as a single polycistronic message.
        4.   A **repressor** is a regulatory protein that blocks transcription (negative control) by binding with the **operator** of the operon.
        5.   An **inducer** is molecule that binds with the repressor and changes its shape so that it can no longer bind with the operator.
        6.   An **activator** is a regulatory protein that enhances transcription (positive control).
    C.   The **lac operon** is an important model for understanding the control of gene expression in bacteria
    D.   **Catabolite repression** turns off certain genes when more readily degradable energy sources such as glucose are available.

7.7   Sensing and responding in response to environmental fluctuations
    A.   **Signal transduction**
        1.   **Two-component regulatory systems** utilize a sensor that recognizes changes outside the cell and then transmits that information to a response regulator.
        2.   Bacteria that utilize **quorum sensing** synthesize a soluble compound, a homoserine lactone, which can move freely in and out of a cell and functions when it reaches a critical concentration.
    B.   **Natural selection**—The expression of some genes changes randomly.

Terms You Should Know

Activator	Induction	P-site
Anticodon	Initiation factor	Quorum sensing
Antiparallel	Intron	Reading frame
A-site	Lagging strand	Regulatory protein
Catabolite repression	Leading strand	Regulatory region
Codon	Minus (-) strand	Release factor
Co-repressor	Monocistronic	Repression
Diauxic growth	Okazaki fragment	Repressor Ribosome-binding site
DNA replication	Open reading frame	Semiconservative replication
Elongation factor	Operator	Sigma factor
E-site	Operon	Start codon
Exon	Phase variation	Stop codon
Gene	Plus (+) strand	Template
Genetic code	Polycistronic	Terminator
Genome	Polyribosome	Transcription
Genomics	Primer	Translation
Inducer	Promoter	

Learning Activities

1. Matching: Match the term on the left with the appropriate definition on the right.

	1. Genome	A.	The triplet of nucleotides in mRNA that codes for a specific amino acid
	2. Gene	B.	RNA that carries the information from DNA to the ribosome
	3. Plus strand of DNA	C.	The complete set of genetic information for a cell
	4. Minus strand of DNA	D.	Codons that do not code for specific amino acids, but stop the production of a protein
	5. DNA polymerase	E.	The sequence of nucleotides in DNA that may code for a single gene product
	6. RNA polymerase	F.	RNA that brings specific amino acids to the ribosome
	7. mRNA	G.	The strand of DNA that is transcribed into messenger RNA
	8. rRNA	H.	An enzyme used in the replication of DNA
	9. tRNA	I.	The science that that deals with the analysis of a DNA sequence
	10. Genomics	J.	The triple of nucleotides in tRNA that identifies the specific amino acid transported by the RNA
	11. Codon	K.	The enzyme used in transcription
	12. Stop codon	L.	The strand of DNA that is the opposite of the strand that is transcribed into messenger RNA
	13. Anticodon	M.	RNA that composes part of the ribosome

2. Using diagrams describe DNA replication.

3. Differentiate among constitutive enzymes, inducible enzymes and repressible enzymes.

4. Indicate which of the following transcription and translation characteristics are prokaryotic and which are eukaryotic.

	mRNA is not processed
	mRNA does not contain introns
	mRNA contains introns, which are removed by splicing
	A cap is added to the 5′ end of mRNA and a poly A tail is added to the 3′ end
	mRNA transcript is transported out of the nucleus so that it can be translated in the cytoplasm
	Translation of mRNA begins as it is being transcribed
	mRNA is often polycistronic
	mRNA is monocistronic
	Translation begins at the first AUG codon
	Translation begins at the first AUG that follows a ribosome-binding site

5. For the following sequence of nucleotides in DNA indicate:

- The order of bases in the minus strand of DNA.
- The order of bases in the plus strand of DNA.
- The order of bases in the mRNA
- Using the codon chart (mRNA) the specific amino acid for each codon specified
- The anticodon for each specific amino acid.

Minus strand of DNA	5′ TACCTACTTCCCGCAACT 3′
Plus strand of DNA	
mRNA (codons)	
Amino acids	
tRNA (anticodons)	

6. Describe the function of the following:

Promoter	
Repressor	
Activator	

7. Explain the difference between an inducible and repressible enzyme.

_____

_____

8. Give the function of the following enzymes.

Enzyme	Function
Primase	
DNA polymerase	
DNA ligase	
DNA gyrase	
RMA polymerase	

9. Given the *lac* operon model

			Operon		
	Control Region		Structural genes		
Regulatory Gene	Promoter	Operator	Z	Y	A

Give the function of each of the following components:

Regulatory Gene	
Promoter	
Operator	
Structural genes	
RNA polymerase	

- If lactose is present in abundant amounts what happens?

_____

_____

- If lactose is used up what happens?

_____

_____

10. Give specific examples and mechanisms of the following types of controls.

Type of control	Example	Mechanism
Negative control—Repression		
Negative control—Induction		
Positive control—Repression		
Positive control—Induction		

11. Compare and contrast the two-component regulatory system and quorum sensing.

12. Explain why antigenic variation occurs.

13. Explain the importance of genomics and what can be learned by searching a computerized database for sequences that have homologies to a newly sequenced gene.

Self Test

1. The best definition of a gene is

    a. a segment of DNA.
    b. a segment of RNA.
    c. three nucleotides that code for a specific amino acid.
    d. a sequence of nucleotides in DNA that codes for a functional product usually a protein.
    e. a sequence of nucleotides in RNA that codes for a functional product usually a protein.

2. Which of the following is made during transcription?

    a. a complementary strand of DNA
    b. a protein molecule
    c. mRNA
    d. tRNA
    e. All of the above

3. Given a segment of minus strand DNA with the nucleotide sequence AGTTACGGTAAT, what would be the sequence on the complementary (+) strand?

    a. 5′ UCAAUGCCAUUA 3′
    b. 5′ ATTACCGTAACT 3′
    c. 5′ TCAATGCCATTA 3′
    d. 5′ TGCAATGGATCG 3′
    e. 5′ AGTTACGGTAAT 3′

4. Using the same segment of DNA as in question 3, what would be the sequence in the mRNA?

    a. 5′ UCAAUGCCAUUA 3′
    b. 5′ AUUACCGUAACU 3′
    c. 5′ TCAATGCCATTA 3′
    d. 5′ TGCAATGGATCG 3′
    e. 5′ AGUUACGGUAAU 3′

5. What type of RNA transports amino acids to the ribosome for protein synthesis?

    a. cRNA
    b. tRNA
    c. mRNA
    d. rRNA
    e. None of the above

6. The backbone of the strand of the DNA molecule is composed of alternating units of

    a. deoxyribose and phosphate.
    b. ribose and phosphate.
    c. deoxyribose and a nitrogenous base.
    d. ribose and a nitrogenous base.

7. The chemical composition of DNA includes all of the following except

    a. adenine.
    b. thymine.
    c. cytosine.
    d. uracil.
    e. guanine.

8. Metabolites that activate gene transcription are known as

    a. repressors.
    b. operators.
    c. inducers.
    d. promoters.
    e. controllers.

9. Anticodons are sequences of nucleotides in what kind of RNA?

    a. mRNA
    b. cRNA
    c. tRNA
    d. rRNA
    e. Anticodons are found in DNA, not RNA.

10. Enzymes that are not subject to regulation by induction or repression are called

    a. promoter enzymes.
    b. repressor enzymes.
    c. constitutive enzymes.
    d. inducible enzymes.
    e. coenzymes.

Thought Questions

1. Using diagrams, illustrate DNA replication, RNA transcription and RNA translation.

2. List and compare the structure and function of the three kinds of RNA.

3. Distinguish between feedback inhibition and end product repression.

Answers to Self Test Questions

1-d, 2-c, 3-c, 4-a, 5-d, 6-a, 7-d, 8-c, 9-c, 10-c

Chapter 8   Bacterial Genetics

Overview

Changes in the genetic makeup of bacteria can occur by mutation or by gene transfer. Mutations are changes in the sequence of nucleotides in the DNA. This chapter discusses the different types of mutations, mutagenic agents, repair mechanisms for damaged DNA, consequences of mutations, and how mutants may be selected. The natural mechanisms of gene transfer, (1) DNA-mediated transformation, (2) transduction and (3) conjugation are also discussed. The nature of plasmids and transposons are described and their roles in gene transfer are presented.

Learning Objectives

After studying the material in this chapter, you should be able to:

1. Define:
   - Mutation
   - Genotype
   - Phenotype
   - Auxotroph
2. List the different types of mutations and explain how they affect cell function.
3. Given the codons (mRNA) for specific amino acids and the sequence of bases in a strand of DNA determine the effect of a:
   - Base substitution mutation
   - Frame shift mutation (addition or deletion)
4. Describe the major physical and chemical causes of mutations.
5. List and describe the methods for the selection of mutant organisms.
6. Describe the following types of DNA repair:
   - Proofreading and mismatch repair
   - Photoreactivation (light repair)
   - Excision repair (dark repair)
   - SOS repair
7. Define and describe the following:
   - Genetic recombination
   - DNA-mediated transformation
   - Transduction
   - Conjugation
8. Differentiate between generalized and specialized transduction.
9. Define plasmid and explain the importance of plasmids.
10. Define or identify
    - F plasmid
    - Resistance plasmid
    - Hfr cell (high frequency of recombination)
    - Transposon

11. Describe the roles of plasmids and transposons in gene transfer.
12. Describe the following barriers to gene transfer:
    • Restriction of DNA
    • Modification enzymes

Key Concepts

1. Mutations are changes in the sequences of nucleotides in DNA.
2. Mutations may be spontaneous or induced by chemicals or radiation.
3. Mutations occur at random.
4. All mutations are expressed in bacteria because they are haploid organisms.
5. Cellular mechanisms exist to repair errors in DNA.
6. While mutations provide a mechanism for natural selection, they can also be detrimental.
7. DNA can be transferred from a donor cell to a recipient cell by (a) DNA-mediated transformation in which "naked" DNA is transferred, (b) transduction in which a bacteriophage transfers the DNA and (c) conjugation in which DNA is transferred by means of a sex pilus.
8. Transposable elements can move DNA within the same organism. This often occurs with antibiotic resistance genes being transferred from chromosome to plasmid.

Summary Outline

8.1   Diversity in Bacteria: the properties of bacteria can change through **mutations**, changes in the chemical structure of DNA, or by changes in gene expression.

Gene mutation

8.2 **Spontaneous mutations** are changes in the nucleotide sequences in DNA, which occur without the addition of agents known to cause mutations.
    2. **Base substitution** usually occurs during DNA replication.
    3. **Point mutations** occur when only one base pair changes.
    4. **Frame shift mutations** involve the **addition** or **deletion** of nucleotides resulting in all genes downstream of a stop codon and in the same operon becoming nonfunctional.
    5. **Transposable elements (transposons)** have the ability to move to any other location in the genome.

8.3 **Induced mutations**
    A.   Chemical **mutagens** alter hydrogen-bonding properties of purines and pyrimidines such that there is an increase in the frequency of mutations.
    1.   **Base analogs** with different hydrogen bonding properties can be incorporated into DNA in place of usual purines and pyrimidines.
    2.   **Intercalating agents** are planar molecules, which insert into the double helix and push nucleotides apart resulting in a frameshift mutation.
    B.   **Transposition**—An insertion mutation results when a transposon integrates into a recipient cell genome.
    C.   Radiation
    1.   **Ultraviolet irradiation** results in the formation of thymine dimers due to the formation of covalent bonds between adjacent thymine molecules on the same strand of DNA.
    2.   **X-rays** cause single strand breaks, double strand breaks, and alterations to the DNA bases.

8.4 **Repair of damaged DNA**
    A.   **Repair of errors in base incorporation**—DNA polymerase has a proofreading function.

    B.   **Repair of thymine dimers**
1. In **light repair** a photoreactivating enzyme breaks the bonds of the thymine dimer, thereby restoring the original molecule.
2. In **excision or dark repair**, the damaged segment is excised by an endonuclease. A new strand is synthesized by DNA polymerase.

    C.   **SOS repair** is a last ditch repair mechanism in which enzymes are induced by damaged DNA resulting in DNA polymerase bypassing the damaged part of the strand, but without proofreading the DNA product.

8.5 Mutations and their consequences
    A.   Genes mutate **independently** of one another.
    B.   Mutations provide a mechanism for **natural selection**.
    C.   Mutations are **expressed rapidly in bacteria** because bacteria are **haploid**.

8.6 Mutant selection
    A.   **Direct selection** involves inoculating cells on a medium on which the mutant but not the parent can grow.
    B.   **Indirect selection** is required when the mutant being sought does not grow on a medium on which the parent grows.
    C.   **Replica plating** involves the simultaneous transfer of all the colonies on one plate to another and the comparison of the growth of individual colonies on both plates.
    D.   **Penicillin enrichment** will increase the proportion of **auxotrophic mutants** in the population.
    E.   **Conditional lethal** mutations cannot be overcome by adding growth factors to the medium.
    F.   The **Ames test** measures whether a suspected carcinogen increases the frequency of reversion; a positive test indicates the suspected carcinogen is a mutagen and therefore a likely carcinogen.

Mechanisms of gene transfer

8.7 DNA-mediated transformation involves the transfer of "naked" DNA.
    A.   **Horizontal (lateral) gene transfer** is the transfer of genes from one bacterium to another.
    B.   **Vertical gene transfer** is transfer of genes from the parent cell to daughter cells in DNA replication.
    C.   **Natural competence** is the ability of a cell to take up DNA
    D.   **Artificial competence**—Cells can be made artificially competent by electroporation, a process whereby the cells are treated with an electric current that makes holes in the cell envelope through which DNA can pass.

8.8   **Transduction** involves the transfer of bacterial DNA by a bacterial virus or bacteriophage.
8.9   **Conjugation** requires cell-to-cell contact by means of a sex pilus encoded on an F plasmid.
8.10   **Plasmids** are extrachromosomal pieces of DNA that code for nonessential information.
    A.   **R plasmids** code for antibiotic resistance
8.11   Genetic Transfer of Virulence Factors
    A.   Genes coding for virulence factors are transferred between bacteria by **transformation**, **transduction**, and **conjugation**.
    B.   Genes coding for virulence factors may be grouped together in **pathogenicity islands**.
8.12   **Gene movement within the same bacterium—transposable elements**
    A.   Transfer of genes to unrelated bacteria
1. Transposable elements often transfer antimicrobial-resistance genes from a chromosome to a plasmid.
2. Plasmids can be transferred by conjugation to unrelated bacteria.

    B.   Structure of transposons – There are different types of transposons that differ in complexity.
8.13   **Barriers to gene transfer**
    A.   Restriction of DNA -DNA entering unrelated bacteria is recognized as foreign and is degraded by a class of deoxyribonucleases.

B. Modification enzymes methylate potential cleavage sites and prevent the DNA from being cleaved.

8.14 Importance of Gene Transfer to Bacteria - **Gene transfer** allows bacteria to survive changing environments by providing cells with a set of new genes.

Terms You Should Know

Alkylating agent
Auxotroph
Bacteriophage
Base analog
Base substitution mutation
Carcinogen
Conjugation
Dark repair (Excision repair)
Direct selection
DNA-mediated transformation
Electroporation
F (fertility) plasmid
Frame-shift mutation
Gene transfer
Generalized transduction
Genetic recombination
Genetics
Genotype

Haploid
Hfr cell
Horizontal gene transfer
Indirect selection
Insertion mutation
Intercalating agent
Mismatch repair
Missense mutation
Mutagen
Mutation
Natural selection
Nonsense mutation
Pathogenicity islands
Penicillin enrichment
Phenotype
Photoreactivation
Plasmid
Plasmid transfer

Point mutation
Proofreading
R (resistance) plasmid
Recombinants
Replica plating
Replicons
Restriction enzymes
Restriction-modification system
Self-transmissible plasmid
Sex pilus
Silent mutation
SOS repair
Specialized transduction
Spontaneous mutation
Stop codon
Transposons
Vertical gene transfer
Wild type

Learning Activities

1. Define transposons (transposable elements) and explain their significance.

2. Describe briefly the following types of mutations:

Base substitution	
Frameshift: addition	
Frameshift: deletion	
Missense	
Nonsense	

3. Given that there is a point mutation (adenine is substituted) * in the minus strand of DNA at the 7th base from the left, finish the table giving the resulting peptide. Circle the changes in the resulting peptide caused by the mutation. What is the importance of these changes?

Minus strand of DNA	3′ TACCTA*C*TTCCCGCAACT 5′
Positive strand of DNA	
mRNA (codons)	
Amino acids	
tRNA (anticodons)	

4. Given that there is a frameshift mutation between the 10th and 11th bases from the left (guanine is added) in the minus strand of DNA, finish the table giving the resulting peptide. Circle the changes in the resulting peptide caused by the mutation. What is the importance of these changes?

Minus strand of DNA	3′ TACCTACTT*CC*CGCAACT 5′
Positive strand of DNA	
mRNA (codons)	
Amino acids	
tRNA	

5. List three possible causes of mutations.

    _____

    _____

    _____

6. Explain what happens in each of the following types of gene repair.

Repair mechanisms	Explanation
Proofreading	
Mismatch repair	
Light repair	
Excision (dark) repair	
SOS repair	

7. Describe using drawings how the Ames test is used to test chemicals for their cancer causing ability.

8. Explain how the following techniques are used in mutant selection.

Technique	Explanation
Direct selection	
Indirect selection	
Replica plating	
Penicillin enrichment of mutants	
Temperature-sensitive (conditional) mutants	

9. Define the following terms:

Genetic recombination	
Conjugation	
DNA-mediated Transformation	
Transduction	
Plasmid	
F plasmid	
Resistance plasmid	
Hfr cell	
Restriction enzyme	

10. Describe the following types of gene transfer.

DNA-mediated transformation	
Transduction	
Conjugation	

11. Although the three types of gene transfer listed in activity 10 differ from one another in how the DNA is delivered, they share common features. List these three common features.

_____

_____

_____

12. List two important barriers to gene transfer.

_____

_____

13. Explain gene transfer and its implications for evolution.

Self Test

1. Resistance to many antibiotics is carried on

   a. R plasmids.
   b. F plasmids.
   c. chromosomes.
   d. sex pili.
   e. antibiotic plasmids.

2. Ultraviolet light damages

   a. cell membranes.
   b. DNA.
   c. RNA.
   d. proteins.
   e. all of the above.

3. Which of the following kinds of mutations will add nucleotides to the DNA strand?

   a. base analogs.
   b. alkylating agents.
   c. intercalating agents.
   d. nitrous acid.
   e. All of the above.

4. Mutations are expressed frequently in bacteria because bacteria are

   a. prokaryotic.
   b. diploid.
   c. haploid.
   d. competent.
   e. None of the above.

5. The segments of DNA that are transferred between cells are called

   a. chromosomes.
   b. transposons.
   c. plasmids.
   d. parts of chromosomes.
   e. Both c and d.

6. The technique in which bacteria from all of the colonies on one plate are simultaneously transferred to the same positions on another plate is called

   a. direct selection.
   b. indirect selection.
   c. replica plating.
   d. transformation.
   e. transduction.

7. The process of DNA repair in which enzymes cut out a damaged section of DNA and other enzymes repair the resulting break is known as

   a. light repair.
   b. photoreactivation.
   c. SOS repair.
   d. excision (dark) repair.
   e. mismatch repair.

8. An organism that requires a growth factor in order to grow is called a(n)

   a. conditional lethal mutant.
   b. temperature-sensitive mutant.
   c. spontaneous mutant.
   d. auxotroph.
   e. wild type.

9. The methods requiring physical contact by which DNA can be transferred from one cell to another is called

   a. transformation.
   b. transduction.
   c. conjugation.
   d. both a and b
   e. both a and c

10. Barriers to DNA transfer include

    a. restriction of DNA.
    b. restriction of RNA.
    c. modification enzymes.
    d. both a and c
    e. both a and b

Thought Questions

1.   Describe the relationship between mutagens and carcinogens.

2.   What is the most common type of mutation that occurs in an auxotroph?

Answers to Self Test Questions

1-a, 2-b, 3-c, 4-c, 5-c, 6-c, 7-d, 8-d, 9-c, 10-d

Chapter 9   Biotechnology and Recombinant DNA

Overview

With the advent of the knowledge of DNA structure and function, it has become possible to manipulate DNA by artificial means forming recombinant DNA. This process is known as genetic engineering. Biotechnology is the use of genetic engineering to solve practical problems and produce useful products. This chapter introduces the topic of recombinant DNA and biotechnology and includes DNA cloning, nucleic acid hybridization, DNA sequencing and the polymerase chain reaction.

Learning Objectives

After studying the material in this chapter, you should be able to:

1. Define:
   - Biotechnology
   - Recombinant DNA
   - Gene cloning
   - Genetic engineering
2. List seven general applications of recombinant DNA technology.
3. List specific proteins made by genetically engineered microorganisms and give their medical uses.
4. Define or identify:
   - Restriction enzyme
   - Recognition sequence
   - Vector
   - Probe
   - Nucleic acid hybridization
   - Restriction fragment length polymorphism (RFLP)
   - Oligonucleotides
   - Reverse transcriptase
   - Introns
   - cDNA
   - Cohesive ends (Sticky ends)
   - DNA ligase
5. Describe how restriction enzymes are used in the production or recombinant DNA.
6. Describe nucleic acid hybridization and list the general procedures that use it.
7. Give the function of the following procedures:
   - Colony blot
   - Southern blot
   - Gel electrophoresis
   - Fluorescence *in situ* hybridization (FISH)
   - Nucleotide array technologies

- DNA sequencing
- Polymerase chain reaction (PCR)

8. Outline the colony blot and describe its use.
9. Outline the Southern blotting procedure and describe its use.
10. Explain the purpose of DNA sequencing.
11. Outline the polymerase chain reaction and describe its use.

Key Points

1. DNA is the hereditary material and it can be studied and manipulated through recombinant DNA technology.
2. Biotechnology is the application of recombinant DNA techniques to practical problems, and often results in the production of useful substances.
3. Restriction enzymes make it possible to cut out specific gene sequences which can then be spliced into foreign DNA creating artificially produced recombinant DNA.
4. To clone DNA fragment, it must be inserted into a vector, creating a recombinant molecule that can replicate in a specific host cell.
5. For DNA to be expressed in a prokaryotic cell, introns must first be removed.
6. The type of vector used to clone eukaryotic DNA depends mainly on the ultimate purpose of the procedure. The vectors are often specific for a particular cell type.
7. Nucleic acid probes can be used to detect a specific sequence of nucleotides in a sample.
8. Colony blots are used to identify colonies that contain a given sequence of nucleotides.
9. Southern blots are used to determine the size restriction fragments that contain a nucleotide sequence of interest and to detect slight variations in nucleotide sequences that are found in different strains of a particular species.
10. Nucleotide array technologies employ a microarray of oligonucleodes each of which act in a manner analogous to a probe.
11. Fluorescent-labeled probe can detect specific nucleotide sequences within intact cells affixed to a microscope slide.
12. By determining the specific sequence of nucleotides in a gene, the protein that it encodes can be fully characterized, and genetic alterations that result in disease can be identified. DNA sequence analysis can also be used to study evolutionary relationships.
13. The polymerase chain reaction (PCR) is an important technique in molecular biology that has made it possible to create many copies of a given region of DNA in a relatively short period of time.
14. Practical applications of DNA technology include (1) production of specific medically important proteins, (2) production of safer and better vaccines, (3) production of pest and/or herbicide-resistant plants, (4) production of plants with improved nutritional value and (5) providing a source of DNA for research.

Summary Outline

9.1   Applications of **genetic engineering**
   A. **Protein production:** Bacteria can be engineered to efficiently produce pharmaceutical proteins, vaccines, and other proteins.
   B. **Providing a source of DNA for study**
      1. Segments of DNA can be cloned into *E. coli* and then used as a source for that sequence for study and further manipulation.

2. Gene function and regulation can be more easily studied in *E. coli* because of the development of systems for the manipulation of DNA.
   C. **Altering organisms** to give them economically useful traits
      1. Transgenic plants can be made using a vector derived from the Ti plasmid of *Agrobacterium tumefaciens*.
      2. Examples include plants that resist pests and herbicides, plants with improved nutritional value, and plants that function as edible vaccines.

9.2   Applications of **nucleic acid hybridization**
   A. **DNA probes** are used to locate a specific nucleotide sequence in a DNA sample.
   B. **Colony blots** are used to identify colonies that contain the DNA sequence of interest.
   C. **Southern blots** are used to the size of the restriction fragments that contain a specific nucleotide sequence.
   D. **Fluorescence *in situ* hybridization (FISH)** – Specific nucleotide sequences within intact cells affixed to a microscope slide are detected by fluorescent-labeled probes.
   E. **Nucleotide array technologies** – Uses a **microarray** that is composed of tens or hundreds of thousands of oligonucleotides. Each oligonucleotide works in a manner similar to that of a probe.

9.3   Applications of **DNA sequencing**
   A. The **nucleotide sequence** of a gene can be used to determine the **amino acid sequence of the protein** for which it codes.
   B. Genetic alterations that occur in some disease can be identified.
   C. Evolutionary relationships can be studied.

9.4   Applications of polymerase **chain reaction**
   A. The **polymerase chain reaction (PCR)** is used to rapidly increase the amount of specific DNA segment in a sample.
   B. The **three-step amplification cycle**
      1. **Double-stranded DNA is denatured**.
      2. **Primers anneal to their complementary sequences**.
      3. **DNA is synthesized**, thus amplifying the target sequence.

9.5   Concerns regarding DNA technologies
   A. Ethical issues have been raised by the advances in **genomics**.
   B. Concerns exist about the introduction of allergens into a food product and adverse effects on the environment.

9.6   Techniques used in genetic engineering
   A. Genetically engineering bacteria
      1. **Cloning into a population of *E. coli*** cells a set of DNA fragments that together make up the entire chromosome of the organism being studied resulting in a DNA library.
      2. **Isolating DNA**
      3. Using restriction enzymes to generate fragments of DNA
      4. Generating a **recombinant DNA molecule**—To enable cloned DNA to replicate in a cell, it is attached to a piece of DNA called a vector, to form a recombinant molecule that is part vector and part cloned DNA.
      5. **Introducing the recombinant DNA** into a new host using transformation or electroporation.
   B. Genetically eukaryotic cells
      1. **Vectors** used to clone DNA into eukaryotic cells are often specific for that cell type.
      2. Introducing DNA into eukaryotic cells
         a. **Ti plasmids** and viruses naturally carry DNA into eukaryotic cells.
         b. Electroporation and a gene gun can be used to move DNA into eukaryotic cells.

9.7 Techniques used in nucleic acid hybridization
   A. **Colony blotting** and **Southern blotting**
      1. **DNA probes**—A probe, which is a labeled single-stranded piece of nucleic acid, is used to locate a specific nucleotide sequence in a DNA sample affixed to a nylon membrane.
      2. **Colony blot**—Colonies are replica-plated on a nylon membrane; a DNA probe is then used to identify colonies that contain the sequence of interest.
      3. **Southern blot**—**Gel electrophoresis** is used to separate DNA fragments according to size and the separated DNA is transferred in place to a nylon membrane. A DNA probe is then added to the membrane to locate specific nucleotide sequences.
   B. **Fluorescence** *in situ* **hybridization** – Samples are treated to preserve the shape of cells, inactivate enzymes, and make the cells permeable.
   C. **Nucleotide array technologies** – **Microarrays** are constructed and combined with the DNA to be studied.
9.8 Techniques used in DNA sequencing
   A. **Dideoxy chain termination method**
      1. A dideoxynucleotide is a nucleotide that lacks the 3'OH and functions as a chain terminator.
      2. The sizes of fragments in a sequencing reaction indicate the positions of the terminating nucleotide base in the synthesized DNA strand.
   B. **Automated DNA sequencing**
      1. Each ddNTP(dideoxynucleotide) is labeled with a different color of fluorescent dye.
      2. The reactions are done in a single tube and run on gel electrophoresis.
      3. A laser detects the color of the band as it runs off the gel.
9.9 Techniques used in **polymerase chain reaction (PCR)**
   A. The three step amplification cycle
      1. **Double-stranded DNA is denatured**.
      2. **Primers anneal to their complementary sequences**.
      3. **DNA is synthesized**, thus amplifying the target sequence.
   B. Generating a discrete-sized fragment
      1. After three cycles of replication a discrete-sized fragment it amplified exponentially.
      2. The size of the amplified fragment is dependent on the positions to which the primers anneal.
   C. The selection of primer pairs determines which portion of the DNA is amplified.

Terms You Should Know

Agarose	Electroporation	Recognition sequence
Biotechnology	Gel electrophoresis	Recombinant DNA
CDNA	Gene cloning	Restriction enzyme
Chain terminators	Gene fusion	Restriction fragment
Cohesive (sticky) ends	Gene gun	Restriction length polymorphisms
Dideoxynucleotide	Genetic engineering	(RFLP)
DNA cloning	Hybridization	Template
DNA cloning	Intron	Ti plasmid
DNA library	Microarray	Transgenic organism
DNA ligase	Oligonucleotide	Vector
DNA polymerase	Probe	

Important Techniques

Identify the following techniques.

Colony blot
Dideoxy chain termination method
DNA sequencing
Fluorescence *in situ* hybridization (FISH)
Nucleic acid hybridization
Nucleotide array technologies
Polymerase chain reaction
Southern blot

Learning Activities

1. List five general applications of genetically engineered bacteria and examples of each.

Category	Example

2. For each of the following kinds of proteins produced by genetically engineered microorganisms, write the medical use.

Protein	Medical Use
Alpha interferon	
Beta interferon	
Deoxyribonuclease	
Factor VIII	
Gamma interferon	
Growth hormone	
Insulin	
Plate derived growth factor	
Streptokinase	
Tissue plasminogen activator	

3. List four groups of transgenic plants and give an application of each.

Group	Example

4. Define:

	Definition
Biotechnology	
Genetic engineering	
Recombinant DNA	
Vector	

4. Describe the function of each of the following in biotechnology:

Enzyme	Function
Restriction enzyme	
DNA ligase	
DNA polymerase	

5. Define DNA probes, and describe how they are used.

6. Describe how a colony blot is done.

7. List the steps in the Southern blotting procedure.

1.
2.
3.
4.
5.
6.

8. List two applications for Southern blotting.

1.
2.

9. Explain the importance of DNA sequencing.

10. Describe the steps in the dideoxy chain termination method of DNA sequencing.

11. List the steps in the polymerase chain reaction.

1.	
2.	
3.	

12. What is the purpose of the polymerase chain reaction?

13. Explain how the polymerase chain reaction generates a discrete sized fragment.

Self Test

1. The genes that are cloned in *Escherichia coli* coding for the protein insulin originally came from

    a. swine.
    b. bovine pancreas.
    c. *Escherichia coli*.
    d. *Salmonella*.
    e. humans.

2. Which of the following can now be produced in microorganisms?

    a. insulin
    b. growth hormone
    c. blood clotting factors
    d. tissue plasminogen activator
    e. all of the above

3. Which of the following molecular techniques allows one to produce million of copies of a given region of DNA in a matter of a few hours?

    a. DNA sequencing
    b. Colony blots
    c. gene cloning
    d. nucleotide array technologies
    e. polymerase chain reaction

4. The most common vectors for gene cloning in *Escherichia coli* are

    a. plasmids
    b. bacteriophages
    c. viruses
    d. both a and c
    e. a, b and c

5. What types of substances are used to cut DNA into specific pieces at defined sites?

    a. repair enzymes
    b. unrestricted endonucleases
    c. restriction enzymes
    d. exonucleases
    e. polymerases

6. The technique by which the introns or intervening sequences of DNA in eukaryotic genes can be removed involves

    a. synthesis of mRNA.
    b. synthesis of cRNA.
    c. synthesis of cDNA.
    d. synthesis of synthetic DNA.
    e. None of the above

7. A process in which two complementary strands from different sources of DNA anneal is called

    a. hybridization.
    b. polymerase chain reaction.
    c. nucleotide sequence analysis.
    d. Both a and c
    e. a, b and c

8. Which of the following technique(s) use(s) nucleic acid hybridization to locate a specific DNA sequence?

    a. colony blot
    b. dideoxy chain termination method
    c. Southern blot
    d. Both a and b
    e. Both a and c

9. Reverse transcriptase is used in which of the following methods?

    a. polymerase chain reaction
    b. removing introns from eukaryotic DNA
    c. removing introns form prokaryotic DNA
    d. Both a and b
    e. a, b and c

10. A probe that is used to identify DNA can be

    a. DNA
    b. RNA
    c. protein
    d. Both a and b
    e. All of the above

Thought Questions

1. Explain why subunit vaccines produced through genetic engineering are safer than vaccines composed of killed or weakened forms of the disease-causing agent.

2. If you wanted to do polymerase chain reaction in your garage laboratory, what would you need?

Answers to Self Test Questions

1-e, 2-e, 3-e, 4-d, 5-c, 6-c, 7-a, 8-e, 9-d, 10-d

Chapter 10   Identification and Classification of Prokaryotes

Overview

A system of classification is necessary to organize the information about microorganisms. The system can be used for identification and to show the relationship between organisms. While phenotypic characteristics such as morphology, cell wall structure, biochemical reactions and staining reactions have traditionally been the basis of classification and identification the advent of biotechnology has enabled the use of genotypic characteristics as well. Methods such as the use of nucleic acid probes and rRNA sequencing are discussed.

Learning Objectives

After studying the material in this chapter, you should be able to:

1. Define
   - Taxonomy
   - Identification
   - Classification
   - Nomenclature
2. Explain why organisms are arranged in taxonomic groups.
3. Give the order of taxonomic groups from the most general to the most specific.
4. Define
   - Domain
   - Genus
   - Species
   - Strain
5. Describe the basis for the identification of prokaryotes using:
   - Phenotypic characteristics
   - Genotypic characteristics
6. Describe the kind of information provided by:
   - Morphological studies
   - Staining reactions
   - Biochemical reactions
   - Serology
   - Phage typing
   - Sequencing rRNA genes
   - Genomic typing
   - DNA hybridization
   - Numerical taxonomy
7. Explain the basis of nucleic acid hybridization and the kind of information it provides.
8. Explain how rRNA sequences can be used in identification and classification.

Key Concepts

1. Taxonomy consists of three interrelated areas: identification, classification and nomenclature.
2. In the clinical laboratory, identification of the genus and species of a microbe that is causing a problem in a patient is more important that understanding its evolutionary relationship with other organisms.
3. Classification is the process of arranging organisms into similar or related groups in order to provide an organized system for identification and study.
4. Identification is the process of characterizing organisms in order to determine their genus and species.
5. Nomenclature is the assignment of names of organisms.
6. Phenotypic and genotypic characteristics can be used to both classify and identify prokaryotes.
7. Identification may be based on morphological and metabolic characteristics or on biochemical composition.
8. Microorganisms can be identified by genotypic characteristics using probes or PCR to detect DNA nucleotide sequences that are unique to a particular organism. Ribosomal genes can be sequenced to identify an organism that cannot be grown in culture.
9. Molecular methods used to determine the relatedness of different prokaryotes for classification purposes include DNA base composition determination, DNA hybridization and comparison of the sequences of 16S rDNA.
10. Molecular techniques can be used to detect subtle genomic differences between strains of microorganisms that are phenotypically identical. These techniques include genomic typing by pulsed-field gel electrophoresis, and ribotyping, phage typing and determination of antibiograms.
11. While historically prokaryotes have been grouped according to phenotypic characteristics, molecular approaches give greater insight into the relatedness of microorganisms.
12. Modern methods of bacterial classification are based on comparative studies of DNA sequences between organisms.

Summary Outline

10.1. Taxonomy
    A.   **Taxonomy** consists of three interrelated areas:
        1. **Identification**
        2. **Classification**
        3. **Nomenclature**
    B.   **Identification** of prokaryotes
        1   **microscopic examination**
        2. **cultural characteristics**
        3. **biochemical tests**
        4. **nucleic acid**
        5. **patient's disease symptoms**
    C.   Taxonomic categories in a hierarchical order include species, genus, order, class, division or phylum, kingdom and domain.
10.2. Using **phenotypic characteristics** to identify prokaryotes
    A.   **Microscopic morphology**: **Size, shape**, and **staining** characteristics of a microorganism
        1. The **Gram stain** is a **differential stain** that distinguishes the Gram-positive and Gram-negative bacteria.
        2. Certain microorganisms have unique identifying characteristics that can be detected by using **special staining procedures**.

B. **Metabolic differences**
1. The use of **selective** and **differential**.
2. Most **biochemical tests** rely on a pH indicator or chemical reaction that results in a color changer when a compound is degraded.
3. Identification using biochemical tests relies on the use of a dichotomous key.

C. **Serology**: The **proteins** and **polysaccharides** that make up a prokaryote are sometimes unique enough to be considered **identifying markers**.

D. Fatty acid analysis: Cellular **fatty acid composition** can be used as an **identifying marker**.

10.3 Using **genotypic characteristics** to identify prokaryotes
A. **Nucleic acid probes** to detect specific DNA sequences
B. **Amplifying specific DNA** using the **polymerase chain reaction**
C. **Sequencing ribosomal RNA genes**

10.4 Characterizing **strain differences**
A. **Biochemical typing:** A strain that has a characteristic **biochemical variation** is called a **biovar** or **biotype**.
B. **Serological typing:** A strain that differs **serologically** from other strains is called a **serovar** or **serotype**.
C. **Genomic typing: Genomic differences** detected by probes
1. **Different isolates** of the same species that have **different restriction fragment length polymorphisms (RFLPs)** are considered **different strains**. **RFLPs** can be detected by **pulsed-field gel electrophoresis** and by **ribotyping**.
D. **Phage typing** based on the patterns of susceptibility to various types of **bacteriophage** can be used to demonstrate strain differences.
E. **Antibiogram**: Antibiotic susceptibility patterns can be used to distinguish strains.

10.5 Difficulties in classifying prokaryotes: **Prokaryotes** have **few differences in size and shape** and **do not undergo sexual reproduction** so it is **difficult to determine their genetic relatedness**.

10.6 **Numerical taxonomy** relies on a battery of **phenotypic characteristics** and classifies bacteria based on their similarity coefficient.

10.7 Using **genotypic characteristics** to classify prokaryotes
A. Based on the comparison of the **nucleotide sequence of the DNA** of different organisms.
B. **DNA base composition** usually expressed as the G + C content. If the G + C content of two organisms differs by more than a few percent, then they are not closely related.
C. **DNA hybridization**: The extent of nucleotide sequence similarity can be determined by measuring how completely single strands of their DNA will anneal to one another.
D. Comparing the **sequences of 16S ribosomal nucleic acid**

Terms You Should Know

Antibiogram	Genomic typing	Order
Biovar (biotype)	Genus	Phage typing
Chromatogram	H antigen	Phylogeny
Class	Identification	Phylum or division
Classification	Kingdom	Serology
Dichotomous key	Nomenclature	Species
Domain	Numerical taxonomy	Strain
Family	O antigen	Taxonomy

Learning Activities

1. Explain why organisms are divided into taxonomic groups.

    _____

    _____

2. Define:

    *   Taxonomy
    *   Identification
    *   Classification
    *   Nomenclature

3. Differentiate between identification and classification.

    _____

    _____

4. Compare the following characteristics in the three Domains:

	Archaea	Bacteria	Eucarya
Presence of peptidoglycan in the cell wall			
Cytoplasmic membrane lipids			
Ribosomes			
Presence of introns in the DNA			
Membrane-bound nucleus			

5. List the order of taxonomic groups from the most general to the most specific.

    Domain, _____ , _____ , _____ , _____ , _____ , _____ , strain

6. Explain the use of the term domain and how that taxonomic group differs from a kingdom.

7. Define

Term	Definition
Genus	
Species	
Strain	

8. List four methods of identification based on determination of the phenotypic characteristics of bacteria. Give an example of each.

Method	Example

9. List three methods of identification based on determination of the molecular characteristics of bacteria.

1.	
2.	
3.	

10. Explain what each of the five methods used to characterize a strain of bacteria specifically examine.

1.	
2.	
3.	
4.	
5.	

11. List and describe briefly two methods used in genomic typing.

Methods	Description

12. List the methods used in determining the relationships between bacteria for the purpose of classification.

1.	
2.	
3.	
4.	
5.	

13. Complete the chart indicating the type of information provided by each type of procedure.

Procedure	Type of information gathered
Morphological studies	
Staining reactions	
Biochemical reactions	
Serology	
Phage typing	
Sequencing rRNA genes	
Genomic typing	
DNA hybridization	
Numerical taxonomy	

14. What is nucleic acid hybridization and what can it determine?

_____

_____

_____

15. Complete the table for the following biochemical tests.

Test	What occurs	Positive reaction
Catalase		
Citrate		
Gelatinase		
Hydrogen sulfide production		
Indole		
Lysine decarboxylase		
Methyl red		
Oxidase		
Phenyalanine deaminase		
Sugar fermentation		
Urease		
Voges-Proskauer		

Self Test

1. Assigning a name to a microorganism isolated from the culture of a patient's specimen is called

    a. taxonomy.
    b. classification.
    c. identification.
    d. nomenclature.
    e. None of the above

2. The ideal system of classification would be based on

    a. structure.
    b. metabolism.
    c. phenotypes.
    d. evolutionary relationships.
    e. PCR.

3. Metabolic differences are based on

   a. cultural characteristics.
   b. differential staining.
   c. biochemical pathways.
   d. phage typing.
   e. serology.

4. Which of the following techniques use genotypic characteristics for the identification of prokaryotes?

   a. nucleic acid probes
   b. amplification of specific DNA sequences by polymerase chain reaction
   c. sequencing of ribosomal RNA genes
   d. Both a and b
   e. a, b and c

5. Which of the following techniques can be used to identify microorganisms that cannot be cultured?

   a. rRNA analysis
   b. biochemical reactions
   c. morphology
   d. phage typing
   e. DNA base composition

6. Which of the following techniques can be used to identify an organism that occurs in very low numbers in a mixed culture?

   a. Amplifying DNA using PCR
   b. Sequencing of rRNA genes
   c. Fatty acid analysis
   d. Identification cannot be done in a mixed culture.
   e. Both a and b are correct.

7. The results of the analysis of 16S ribosomal RNA demonstrate that

   a. archaea are a separate group from the prokaryotes.
   b. prokaryotes and eukaryotes actually belong to the same group.
   c. bacteria are closely related to certain plants.
   d. archaea are a subgroup within and closely related to the prokaryotes.
   e. there are many as a dozen different groups of organisms very distinct from each other.

8. The GC content of two organisms was found to be 23% and 24%. Which of the following statements is true?

   a. The two organisms are closely related to each other.
   b. The two organisms are not closely related to each other.
   c. The two organisms are closely related only if they belong to the same taxonomic group.
   d. The similarity in GC contents may be coincidental and the organisms may not be related.
   e. Two of the above are correct answers.

9. When identifying an unknown organism,

   a. a few selected tests are usually preferred for an accurate identification.
   b. culturing the organisms is necessary before identification can be made.
   c. the more tests that can be performed the better.
   d. only the techniques of molecular biology are accepted as valid.
   e. Two of the above are correct.

10. Which of the following procedures is the most accurate and reliable for identifying evolutionary relationships of organisms?

   a. Sequencing of 16S ribosomal nucleic acid
   b. Determination of phenotypic characteristics
   c. Fatty acid analysis
   d. Amplifying DNA using PCR
   e. All of the above.

Thought Questions

1. Based on a comparison of the 16S ribosomal RNA, what are the three distinct Domains and how do their cell types differ?

2. Serological testing is based on antigen-antibodies interaction and phage typing is based on the fact that bacteria infected with bacterial viruses (phages) will show the same pattern of infection if the bacteria had a common source. Explain how these can be used to trace the source of an infection.

Answers to Self Test Questions

1-c, 2-d, 3-c, 4-e, 5-a, 6-a, 7-a, 8-e, 9-c, 10-a

Chapter 11   The Diversity of Prokaryotic Organisms

Overview

This chapter presents a survey of the prokaryotic microorganisms with an emphasis on 1) patterns of metabolism and 2) ecophysiology. These organisms are extremely diverse and exist under a great variety of conditions. Some of these organisms are beneficial to human life, others have no direct effect and still others cause disease.

Learning Objectives

After studying the material in this chapter, you should be able to:

1. Define
   - Anaerobe
   - Chemolithotrophs
   - Chemoorganotrophs
2. List the types of prokaryotic microorganisms that are found in the following groups:
   - Anaerobic chemolithotrophs
   - Anaerobic chemoorganotrophs—anaerobic respiration
   - Anaerobic chemoorganotrophs—fermentative
3. Define:
   - Anoxygenic phototrophs
   - Oxygenic phototrophs
4. List and describe the types of prokaryotic microorganisms that are found the in the following groups:
   - Anoxygenic phototrophs
   - Oxygenic phototrophs
5. Define:
   - Aerobic chemolithotrophs
   - Aerobic chemoheterotrophs
6. List and describe the kinds of microorganisms included in
   - Aerobic chemolithotrophs
   - Aerobic chemoheterotrophs
7. List the kinds of bacteria that:
   - Form endospores
   - Produce cysts
   - Produce fruiting bodies
   - Form conidia at the end of hyphae
8. List and describe the kinds of prokaryotic microorganisms that live in an aquatic environment.
9. List and describe the kinds of prokaryotic microorganisms that use animals as their habitat.
10. List and describe the kinds of prokaryotic microorganisms that live under extreme conditions.

Key Concepts

1. Prokaryotic organisms are extremely diverse and live under a wide variety of conditions.
2. Some prokaryotic organisms can derive energy from the oxidation of compounds (chemotrophs) while others derive energy from sunlight (phototrophs).
3. Chemotrophs can be divided into two groups: those that obtain energy by oxidizing organic chemicals (chemoorganotrophs) and those that obtain energy by oxidizing inorganic compounds (chemolithotrophs).
4. Methanogens oxidize hydrogen gas, using Co_2 as a terminal electron acceptor, to generate methane.
5. Sulfur- and sulfate-reducing bacteria oxidize organic compounds, with sulfur compounds serving as terminal electron acceptors, to generate hydrogen sulfide (H_2S).
6. The lactic acid bacteria oxidize organic compounds, with an organic compound serving as a terminal electron acceptor.
7. Phototrophs harvest energy from sunlight and can be divided into two groups: anoxygenic and oxygenic. Anoxygenic phototrophs, unlike oxygenic phototrophs, do not generate oxygen because they do not use water as a source of electrons.
8. Obligate aerobes can generate energy only through aerobic respiration. Facultative anaerobes prefer to use aerobic respiration to generate energy, but can use fermentative metabolism if oxygen is unavailable.
9. *Bacillus* and *Clostridium* species produce endospores. These dormant forms are very resistant to heat and drying and enable these bacteria to survive adverse conditions.
10. *Agrobacterium* and *Rhizobium* derive nutrients from plants, although the former are plant pathogens and the latter benefit the plant.
11. Prokaryotic microorganisms have a variety of mechanisms that help them to live in aquatic habitats. These include production of sheaths; production of prosthecae, which are extensions that maximize the absorptive surface area; bioluminescence; preying upon other bacteria; using unusual methods of locomotion such as axial filaments or magnetic crystals to move to more desirable locations; and storage granules.
12. Some prokaryotic organisms use animals, including humans, as their habitat; these include *Staphylococcus* species, that live under dry, salty conditions; *Bacteroides, Bifidobacterium* species, *Campylobacter* species and *Helicobacter* species that live in the gastrointestinal tract; *Neisseria* species, mycoplasma and spirochetes that inhabit other mucous membranes; and *Rickettsia, Orientia, Ehrlichia, Coxiella and Chlamydia* species which are obligate intracellular parasites.
13. Archaea inhabit extreme environments that include conditions of excess salinity, heat, acidity and alkalinity.

Summary Outline

11.1 **Anaerobic Chemotrophs**
    A. **Anaerobic organisms** use **terminal electron** acceptors other than O_2
    B. **Anaerobic chemolithotrophs**
        1. Oxidize inorganic compound such as hydrogen to obtain energy
        2. CO_2 is the terminal electron acceptor
        3. Example: Methanogens (Domain Archae)
    C. **Anaerobic chemoorganotrophs**—anaerobic respiration
        1. Oxidize organic compounds such as glucose to obtain energy
        2. The terminal electron acceptor is an organic compound other than O_2
        3. Example: *Desilfovibrio*

   D. **Anaerobic chemoorganotrophs**—fermentation
       1. The **end products of fermentation** include a variety of **acids** and **gases** that
       2. are generally characteristic for a given species.
       3. *Clostridium species* are **Gram-positive rods**
       4. The **lactic acid bacteria** are a group of **Gram-positive organisms** that produce lactic
           acid as their primary fermentation end-products.

11.2 Anoxygenic phototrophs
   A. **Phylogenetically diverse group** of bacteria that harvest the energy of sunlight, using
       photosynthesis to synthesize organic materials.
   B. The purple bacteria
       1. The **purple bacteria are Gram-negative organisms** that appear red, orange or purple;
           the photosynthetic apparatus is contained within the cytoplasmic membrane.
       2. The **purple sulfur bacteria** preferentially use **sulfur** as a **source of reducing power**.
       3. The **purple nonsulfur bacteria** preferentially use **organic molecules** as a **source of
           reducing power**.
   C. The **green bacteria**
       1. The **green bacteria are Gram-negative organisms** that are typically green or brownish
           in color. Their light harvesting pigments are located in structures called **chlorosomes**.
       2. The **green sulfur bacteria** use **hydrogen sulfide** as a **source of reducing power**.
       3. The **green nonsulfur bacteria** are characterized by their **filamentous growth**;
           metabolically, they resemble the purple nonsulfur bacteria.
       4. Other **anoxygenic phototrophs** include a Gram-positive rod that forms endospores.

11.3 **Oxygenic phototrophs**
   A. The **cyanobacteria** are a diverse group of **Gram-negative bacteria** that are essential
       primary producers; unlike eukaryotic photosynthesizers, they can **fix nitrogen**.
   B. Genetic evidence indicates that **chloroplasts** of plants and algae evolved from a species of
       cyanobacteria.
   C. **Nitrogen-fixing cyanobacteria** provide an available source of both **carbon** and **nitrogen**.
   D. **Filamentous cyanobacteria** may be involved in maintaining the structure and productivity of
       some soils.
   E. **Some species** of cyanobacteria **produce toxins** that can be deadly to animals that ingest
       heavily contaminated water.

11.4 **Aerobic chemolithotrophs**
   A. **Aerobic chemolithotrophs** generate energy by oxidizing reduced **inorganic compounds**
       using **O_2 as a terminal electron acceptor**.
   B. **Sulfur-oxidizing bacteria** are **Gram-negative rods or spirals**, sometimes growing in
       filaments.
   C. The **filamentous sulfur-oxidizers** *Beggiatoa* and *Thiothrix* live in sulfur springs, sewage-
       polluted waters, and on the surface of marine and freshwater sediments.
   D. Nitrifiers—**Ammonia oxidizers** convert **ammonia to nitrite** and include *Nitrosomonas* and
       Nitrosococcus; **nitrite oxidizers** oxidize **nitrite to nitrate** and include *Nitrobacter* and
       Nitrococcus.
   E. **Hydrogen-oxidizing bacteria** are **thermophilic bacteria** that are thought to be among the
       earliest bacterial forms.

11.5 **Aerobic chemoorganotrophs** oxidize **organic compounds** for energy using **O_2 as a terminal
    electron acceptor**.
   A. **Obligate aerobes** generate energy exclusively by respiration.
       1. *Micrococcus* species are **Gram-positive cocci** found in soil and on dust particles,
           inanimate objects, and skin.
       2. *Mycobacterium* species are **acid-fast**.

3. *Pseudomonas* species are **Gram-negative rod-shaped bacteria** that are widespread in nature and have extremely diverse metabolic capabilities.
4. *Thermus aquaticus* is the source of *Taq* **polymerase**, which is an essential component in the **polymerase chain reaction**.
5. *Deinococcus radiodurans* can survive high doses of gamma radiation.

B. **Facultative anaerobes**

1. *Corynebacteium* species are **Gram-positive pleiomorphic rod-shaped organisms** that commonly inhabit the soil, water and the surface of plants.
2. Members of the family **Enterobacteriaceae** are **Gram-negative rods** that typically inhabit the intestinal tract of animals, although some reside in rich soil. **Enterics** that **ferment lactose** are included in the group called **coliforms** and are used as indicators of fecal pollution.

11.6 Ecophysiology: Thriving in terrestrial environments

A. **Bacteria** that **form a resting stage**

1. **Endospores** are most resistant to environmental extremes.
2. **Endospore-forming** genera include *Bacillus* and *Clostridium*.
3. *Azotobacter* species are **Gram-negative pleiomorphic rods** that form a resting cell called a **cyst** and are notable for their ability to **fix nitrogen** under aerobic conditions.
4. The **myxobacteria** aggregate to form a **fruiting body** when nutrients are exhausted; within the fruiting body cells differentiate to form a **dormant microcyst**.
5. *Streptomyces* species are **Gram-positive bacteria** that resemble fungi in their pattern of growth; they form chains of conidia at the end of hyphae. Many species **naturally produce antibiotics**.

B. **Bacteria that associate with plants**

1. *Agrobacterium* **species** cause the plant disease crown gall.
2. *Rhizobium* species reside as **endosymbionts** that fix nitrogen and reside within cells in nodules formed on the roots of legumes.

11.7 Ecophysiology: Thriving in aquatic environments

A. **Sheathed bacteria** form chains of cells encased in a tube; the sheath **enables cells to attach to solid objects** in favorable habitats while sheltering them from attack by predators.

B. **Prosthecate bacteria**

1. Examples include *Caulobacter* species and *Hyphomicrobium* species.

C. **Bacteria that derive nutrients from other organisms**

1. *Bdellovibrio* species are highly **motile Gram-negative rods** that prey on other bacteria.
2. Certain species of **bioluminescent bacteria** of the Gram-negative genera *Photobacterium* and *Vibrio* establish a symbiotic relationship with specific types of squid and fish.
3. *Legionella* species often reside within protozoa and can **cause respiratory disease** when inhaled in aerosolized droplets.

D. **Bacteria that move by unusual mechanisms**

1. **Spirochetes** are a group of **Gram-negative spiral-shaped bacteria** that move by means of an **axial filament**.
   a. *Spirochaeta* thrive in mud and anaerobic waters.
   b. *Leptospira interrogans* causes **leptospirosis**.
   c. *Treponema pallidum* causes **syphilis**.
2. **Magnetotatic bacteria** such as *Magnetospirillum magnetotacticum* contain a string of **magnetic crystals** that enable them to move up or down in water or sediments to the microaerophilic niches they require.

E. **Bacteria that form storage granules**
1. *Spirillum volutans* stores **polyphosphate granules**.
2. *Thiomargarita namibiensis* **stores granules of sulfur** and has a nitrate-containing vacuole.

11.8 Ecophysiology: Animals as habitats

A. **Bacteria that inhabit the skin**
1. *Staphylococcus* **species** are **Gram-positive cocci** that are facultative anaerobes.
2. *S. epidermidis* is part of the normal flora of the skin.
3. *S. aureus* causes a variety of diseases including **skin and wound infections**, as well as **food poisoning**.

B. **Bacteria that inhabit mucous membranes**
1. *Bacteroides* **species** are strictly **anaerobic Gram-negative** rods that **inhabit the mouth, intestinal tract,** and **genital tract** of humans and other animals.
2. *Bifidobacterium* **species** are **irregular Gram-positive rods** that reside primarily in the **intestinal tract** of animals.
3. *Campylobacter* and *Helicobacter* **species** are **microaerophilic, curved Gram-negative rods**.
   a. *C. jejuuni* causes **diarrheal disease** in humans.
   b. *H. pylori* causes **stomach ulcers**.
4. *Haemophilus* **species** are **Gram-negative coccobacilli** that require compounds found in blood for growth.
   a. *H. influenzae* causes a variety of diseases, primarily in children
   b. *H. ducreyi* causes **chancroid**.
5. *Neisseria* **species** are **Gram-negative diplococci** that are nutritionally **fastidious, obligate aerobes** that grow in the **oral cavity** and **genital tract**.
   a. *N. meningitidis* causes **meningitis**.
   b. *N. gonorrhoeae* causes **gonorrhea**.
6. *Mycoplasma* **species lack a cell wall**; they often have sterols in their membrane that provides strength and rigidity. *M. pneumoniae* causes a form of **pneumonia**.
7. *Treponema* and *Borrelia* **species** are **spirochetes** that typically inhabit mucous membranes and body fluids of humans and other animals.
   a. *T. pallidum* causes **syphilis**.
   b. *B. recurrentis* and *B. hermsii* cause **relapsing fever**.
   c. *B. burgdorferi* causes **Lyme disease**.

C. **Obligate intracellular parasites** are unable to reproduce outside a host cell; most have lost the ability to synthesize substances needed for extracellular growth.
1. Species of *Rickettsia*, Orientia and *Ehrlichia* are tiny **Gram-negative rods** that are spread when a blood-sucking **arthropod** transfers bacteria during a blood meal.
   a. *R. rickettsi* causes **Rocky Mountain spotted fever**.
   b. *R. prowazekii* causes **epidemic typhus**.
   c. *O. tsutsugamushi* causes **scrub typhus**.
   d. *E. chaffeenis* causes human **ehrlichiosis**.
2. *Coxiella burnetti* is a **Gram-negative rod** that survives well outside the host due to the production of **spore-like structures**.
3. *Chlamydia* **species** are transmitted directly from person to person.

11.9 **Ecophysiology**: The **Archaea thrive in extreme environments**
  A. **Extreme halophiles** are found in salt lakes, soda lakes, and brines used for curing fish; they can grow well in saturated salt solutions. They include *Halobacterium, Halorubrum, Natronobacterium* and *Natronococcus*.
  B. **Extreme thermophiles**
    1. *Methanothermus* and *Methanopyrus* are **hyperthermophiles** that **generate methane**.
    2. **Sulfur- and sulfate-reducing hyperthermophiles** are **obligate anaerobes** that use **sulfur** or sulfate as a **terminal electron acceptor**. They include *Thermococcus, Archaeoglobus, Thermoproteus, Pyrodictium* and *Pyrolobus*.
    3. **Sulfur-oxidizing hyperthermophiles** oxidize sulfur compounds, using O_2 **as a terminal electron acceptor**, to **generate sulfuric acid**. They are exemplified by the genus *Sulfolobus*, which is an obligate aerobe.
  C. **Thermophilic extreme acidophiles** include *Thermoplasma* and *Picrophilus* **species**.

Terms You Should Know

Ammonia oxidizer
Anoxygenic phototroph
Bioluminescence
Chemolithotroph
Chemoorganotroph
Chemotroph
Coliforms
Diphtheroids
Endospore

Endosymbiont
Enterics
Facultative anaerobe
Legume
Metachromatic granules
Methanogen
Nitrification
Nitrifier
Nitrite oxidizer

Nitrogen fixation
Opportunistic pathogen
Oxygenic phototroph
Phototroph
Primary producer
Saprophytes
Terminal electron acceptor

Learning Activities

1. Define the following terms.

Term	Definition
Chemotroph	
Chemolithotroph	
Chemoorganotroph	
Phototroph	
Anaerobe	
Aerobe	
Anoxygenic	
Oxygenic	

2.   Name the types of microorganisms that belong to the following groups:

Group	Types of microorganisms
Anaerobic chemolithotrophs	
Anaerobic chemoorganotrophs—anaerobic respiration	
Anaerobic chemoorganotrophs—fermentation	
Anoxygenic phototrophs	
Oxygenic phototrophs	
Aerobic chemolithotrophs	
Aerobic chemoorganotrophs	

3.   List the kinds of prokaryotic microorganisms that live under the following conditions:

Conditions	Microorganisms
Terrestrial environments: Form endospores	
Terrestrial environments: Form resting forms other than endospores	
Terrestrial environments: Bacteria associated with plants	
Aquatic environments: Sheathed bacteria	
Aquatic environments: Prosthecate bacteria	
Aquatic environments: Bacteria that derive nutrients from other organisms	
Aquatic environments: Bacteria that move by unusual mechanisms	
Aquatic environments: Bacteria that form storage granules	
Animals as habitats: Bacteria that inhabit the skin	
Animals as habitats: Bacteria that inhabit mucous membranes	
Animals as habitats: Obligate intracellular parasites	
Archaea: Extreme halophiles	
Archaea: Extreme thermophiles	
Archaea: Thermophilic extreme acidophiles	

4. Match the following: (Answers may be used more than once.)

	Purple sulfur bacteria	a. Use organic compounds
	Purple nonsulfur bacteria	b. Use sulfur compounds as an energy source
	Green sulfur bacteria	c. Use hydrogen sulfide as a source of electrons
	Green nonsulfur bacteria	d. Photosynthetic and generate O_2
	Cyanobacteria	e. Oxidize nitrogen compounds such as ammonium or nitrite
	Sulfur oxidizers	f. Many species can fix nitrogen
	Nitrifiers	

5. Match the following:

	1. Facultatively anaerobic Gram-negative rods	A. Gram-negative spiral organisms with axial filaments that are difficult to culture in vitro
	2. Enteric organisms	B. Lives in an aquatic environment
	3. *Bacteroides*	C. Intracellular parasite that requires an arthropod vector for transmission
	4. Spirochetes	D. Found primarily in the intestines of humans
	5. *Rickettsia*	E. *Salmonella, Shigella, Escherichia, and Enterobacter*
	6. *Legionella*	F. Acid-fast organism
	7. *Mycoplasma*	G. Gram-negative anaerobic bacteria
	8. *Staphylococcus*	H. Catalase-negative
	9. *Streptococcus*	I. Lacks a cell wall
	10. *Mycobacterium* species	J. Catalase-positive

Self Test

1. Prokaryotic microorganisms that use chemicals for an energy source and produce lactic acid as a terminal electron acceptor belong to which of the following groups?

    a. anaerobic chemolithotrophs—methanogens
    b. anaerobic chemoorganotrophs—anaerobic respiration
    c. anaerobic chemoorganotrophs—fermentative
    d. anoxygenic phototrophs
    e. oxygenic phototrophs

2. Prokaryotic microorganisms that use light as an energy source and hydrogen sulfide or organic compounds as a source of electrons for reduction belong to which of the following groups?

    a. anaerobic chemolithotrophs—methanogens
    b. anaerobic chemoorganotrophs—anaerobic respiration
    c. anaerobic chemoorganotrophs—fermentative
    d. anoxygenic phototrophs
    e. oxygenic phototrophs

3. Gram-positive rods that form endospores belong to which of the following genera?

   a. *Clostridium*
   b. *Streptococcus*
   c. *Bacillus*
   d. *Propionibacterium*
   e. Both a and c

4. Gram-positive pleiomorphic rods that produce propionic acid and ferment lactic acid belong to which of the following genera?

   a. *Clostridium*
   b. *Streptococcus*
   c. *Bacillus*
   d. *Propionibacterium*
   e. *Lactococcus*

5. Which of the following bacteria are anoxygenic phototrophs?

   a. purple sulfur bacteria
   b. purple nonsulfur bacteria
   c. green sulfur bacteria
   d. green nonsulfur bacteria
   e. All of the above

6. Which of the following bacteria are obligate aerobic, Gram-positive, acid-fast rods?

   a. *Micrococcus*
   b. *Mycobacterium*
   c. *Mycoplasma*
   d. *Pseudomonas*
   e. *Corynebacterium*

7. Which of the following bacteria are Gram-positive pleiomorphic rods containing metachromatic granules?

   a. *Micrococcus*
   b. *Mycobacterium*
   c. *Mycoplasma*
   d. *Pseudomonas*
   e. *Corynebacterium*

8. Which of the following are obligate intracellular parasites?

   a. *Chlamydia*
   b. *Mycoplasma*
   c. *Rickettsia*
   d. *Campylobacter*
   e. Both a and c

9. Which of the following are adaptations that will help bacteria survive in an aquatic environment?

   a. sheaths
   b. prosthecae
   c. storage granules
   d. endospores
   e. a, b, and c are correct.

10. Which of the following are bacteria that inhabit the skin?

    a. *Staphylococcus*
    b. *Streptococcus*
    c. *Enterococcus*
    d. *Lactococcus*
    e. All of the above.

Thought Questions

1. *Rickettia* and *Chlamydia* are both obligate intracellular parasites. How do they differ?

2. Chemolithotrophs vary in energy sources. List three of these energy sources.

Answers to Self Test Questions

1-c, 2-d, 3-e, 4-d, 5-e, 6-b, 7-e, 8-e, 9-e, 10-a

Chapter 12   The Eukaryotic Members of the Microbial World

Overview

Microbiology includes some eukaryotic organisms as well as bacteria and viruses. In this chapter the eukaryotic members of the microbial world are described including algae, protozoa, fungi, slime molds, water molds as well as multicellular parasites. Algae, slime molds and water molds do not directly cause human disease. Protozoa, fungi, arthropods and helminths do cause some important human diseases as well as causing damage to plants and goods. Algae are key primary producers. Fungi are important both ecologically and economically. Arthropod vectors are discussed.

Learning Objectives

After studying the material in this chapter, you should be able to:

1. Describe the characteristics and roles of algae.
2. Explain why algae are important to human existence.
3. Describe the structure of macroscopic algae.
4. Describe the role of algae in paralytic shellfish poisoning.
5. List the characteristics of protozoans.
6. Name the diseases for which the following are causative agents:
   - *Toxoplasma gondii*
   - *Plasmodium*
   - *Trypanosoma*
   - *Trichomonas vaginalis*
   - *Entamoeba histolytica*
   - *Balantidium coli*
   - *Cryptosporidium parvum*
7. List the kinds of organisms included in the fungi.
8. List the characteristics of the four groups of true fungi and describe the roles of fungi.
9. Define or identify dimorphic fungi.
10. Describe the growth requirements of fungi.
11. Describe reproduction in fungi.
12. Explain how fungi cause human disease.
13. Explain why fungi are of economic importance.
14. Describe slime molds and water molds and explain why they are no longer considered to be fungi.
15. Define and give examples:
    - Vector
    - Mechanical vector
    - Biological vector
16. Give the infectious agent transmitted by the following arthropod vectors:
    - Mosquito
    - Flea
    - Louse
    - Tick

17. Describe the role of mites in scabies.
18. List and describe the three major groups of helminths.
19. Explain how helminthic diseases are transmitted to humans

Key Concepts

1. Some eukaryotic organisms are included in microbiology because they are microscopic and they are studied with techniques similar to those used to study bacteria and archaea. These include *Gonyaulax* (algae), *Entamoeba* and *Plasmodium* (protozoa), *Candida albicans* and *Penicillium* (fungi), *Phytophthora infestans* (water mold), *Anopheles* (insect), *Dermacentor* (arachnid), and *Enterobius vermicularis* and *Taenia solium* (helminths)

2. Algae, protozoa, fungi, slime molds and water molds and multicellular parasites, arthropods and helminths are eukaryotic organisms and thus are composed of eukaryotic cells.

3. Algae are important because they are primary producers and because they produce a large proportion of the molecular oxygen (O_2) in the atmosphere.

4. Algae do not directly cause human disease, but some produce toxins that produce harmful effects when ingested by humans. This occurs when fish or shellfish feed on algae and accumulate toxins.

5. Protozoa are unicellular eukaryotic microorganisms that lack photosynthetic capability, are usually motile during some stage of their life cycles and often reproduce by asexual fission.

6. Protozoa are an important part of the zooplankton and the food chain: some protozoa cause very significant diseases including malaria.

7. Fungi are eukaryotic organisms that are crucial in the decomposition or organic materials.

8. Fungi cause significant damage to crops, stored foods, and other goods, and may cause disease.

9. There are three kinds of fungi-like organisms, the acellular slime molds, the cellular slime molds, and the water molds; they are examples of convergent evolution. Some such as *Phytophthora infestans* cause serious diseases of food crops.

10. Arthropods are joint-legged animals that may act as vectors to transmit various microbial agents that cause disease.

11. Mechanical vectors pick up pathogens on their bodies from contaminated material and carry them to food consumed by humans; biological vectors host part of the life cycle of the pathogen.

12. Helminths are multicellular animals, including pinworms and tapeworms, which cause human disease.

Summary Outline

Classification of the Eucarya
    A.   Cell structure in Eucarya is different from that seen in Bacteria or Archaea.
    B.   Use of the terms algae, fungi and protozoa are not accurate classification terms when you consider the rRNA sequences of these organisms.
12.1. **Algae** are a diverse group of photosynthetic organisms that contain chlorophyll.
    A.   **Classification of algae** is based on their major **photosynthetic pigments**. Organisms are placed on the phylogenetic tree according to **rRNA sequences**.
    B.   **Algae** are found in **fresh** and **salt water** as well as **soil; unicellular algae** make up part of the **phytoplankton**.

C. **Structure** of algae
   1. **Microscopic** or **macroscopic**.
   2. **Cell walls** are made of **cellulose** and materials such as agar and carrageenan.
   3. They have membrane **bound organelles** including a nucleus, chloroplasts and mitochondria.
D. **Algal reproduction** is **asexual** or sexual.
E. **Paralytic shellfish poisoning** is caused by toxins that are ingested by fish and shellfish.

12.2 **Protozoa** are **microscopic, unicellular organisms** that lack chlorophyll, are motile during a least one stage in their development and reproduce most often by binary fission.
   A. **Classification** of protozoa is based on **rRNA**.
      1. Protozoa have traditionally been grouped based on their mode of locomotion.
      2. **Sarcomastigophora** include
         a) **Mastigophora**—the **flagellated protozoa**
         b) **Sarcodina**—move by means of **pseudopodia**
      3. **Ciliophora** move by means of **cilia**.
      4. **Apicomplexa (sporozoa)** include *Plasmodium* **sp.**, the cause of **malaria**.
      5. **Microsporidia**, an intracellular protozoan, causes disease in immunocompromised individuals.
   B. **Protozoa** are usually **free-living** and found in **marine and fresh water** as well as **terrestrial environments**. They are important **decomposers** and are an important part of the **food chain**.
   C. **Structure** of protozoa
      1. Protozoa **lack a cell wall** but most maintain a definite shape using the underlying ectoplasm.
      2. **Life cycles** are often **complex** and include more than one habitat.
      3. **Protozoa** feed by either **phagocytosis** or **pinocytosis**.
   D. Protozoan **reproduction** is often by **binary fission**; some reproduce by **multiple fission** or **schizogny**.
   E. Protozoa cause some serious disease such as **malaria, sleeping sickness, toxoplasmosis** and **vaginitis**.

12.3 **Fungi** can cause serious disease, primarily in plants, but they also produce useful food products. They include yeast, molds and mushrooms.
   A. **Classification** of fungi includes four groups of true fungi
      1. **Zygomycetes**
      2. **Ascomycetes**
      3. **Basidiomycetes**
      4. **Deuteromycetes** or **Fungi Imperfecti**
      5. Chytridiomycetes are close relatives.
   B. Structure
      1. **Fungal filaments** are called **hyphae** and a group of hyphae is called a **mycelium**.
      2. **Dimorphic fungi** can grow either as a **single cell (yeast)** or a **mycelia**.
   C. **Fungi inhabit** just about every ecological niche and can spoil a large variety of food materials because they can grow in high concentrations of sugar, salt and acid.
      1. Fungi can be found in **moist environments** at temperatures from **-6°C to 50°C** and **pH from 2.2 to 9.6.**
      2. Fungi are heterotrophs with enzymes that can degrade most organic materials.
   D. **Fungal disease** in humans
      1. Fungi may produce an **allergic reaction**.
      2. They may produce a **toxin** that can make humans ill such as **ergot, poisonous mushrooms,** or **aflatoxin**.

3.   They cause mycoses such as
    a)   **Histoplasmosis**
    b)   **Coccidiodomycosis**
    c)   **Candidiasis**
E.   **Symbiotic relationships** between fungi and other organisms
    1.   **Lichens** result from an association of a fungus with a photosynthetic organism such as an alga or a cyanobacterium.
    2.   **Mycorrhizas** are the result of an intimate association of a fungus and the roots of a plant.
F.   **Economic importance** of fungi
    1.   The yeast *Saccharomyces* is used in the production of **beer, wine and bread**.
    2.   **Penicillin** and other fungi **synthesize antibiotics**.
    3.   Fungi **spoil** many **food** products.
    4.   Fungi **cause diseases of plants** such as Dutch elm disease and wheat rust.
    5.   Fungi have been **useful tools in genetic and biochemical studies**.

12.4   **Acellular and cellular slime molds** are important links in the **terrestrial food chain**. **Oomycetes**, also known as water molds, cause some serious **diseases of plants**.

12.5   **Multicellular parasites**: arthropods and helminths
A.   **Arthropods**
    1.   Arthropods act as **vectors for disease**.
    2.   **Mosquitoes** spread diseases such as **malaria** by picking up disease-causing organisms when the mosquito bites, and later injecting organisms into subsequent organisms that it bites.
    3.   **Fleas** transmit disease such as **plague, lice** can transmit **trench fever, epidemic typhus** and **relapsing fever**.
    4.   **Ticks** are implicated in **Rocky Mountain spotted fever** and **Lyme disease**.
    5.   **Mites** cause **scabies**, and **dust mites** are responsible for **allergies** and **asthma**.
B.   **Helminths**
    1.   Most **nematodes** or **roundworms** are free-living, but they may cause serious disease such as **pinworm disease, whipworm disease, hookworm disease** and **ascariasis**.
    2.   **Cestodes** are **tapeworms** with segmented bodies and hooks to attach to the wall of the intestine. Most tapeworm infections occur in persons who eat uncooked or undercooked meats; some tapeworms are acquired by ingesting fleas infected with dog or cat tapeworms.
    3.   **Trematodes**, or **flukes**, often have complicated life cycles that necessarily involve more than one host.
    4.   *Schistosoma mansonii* cercaria can **penetrate the skin** of persons wading in infected waters and cause serious disease.

Terms You Should Know

Acellular slime mold
Alfatoxins
Algology
Arthropods
Biological vector
Bladder (float)
Blade
Candidiasis
Cellular slime mold
Cercaria
Cestode
Coccidioidomycosis
Cyst
Dimorphic fungi
Diploid
Ergot
Foraminifera

Gametes
Germ tube
Haploid
Haustoria
Helminths
Histoplasmosis
Holdfast
Hyphae
Lichen
Mechanical vector
Meiosis
Molds
Mycelium
Mycology
Mycorrhiza
Mycoses
Nematode

Neurotoxin
Oomycetes (water mold)
Phytoplankton
Plasmodium
Protozooloogy
Pseudopodia
Rhizoids
Schizogony
Stipe
Trematode
Trophozoite
Vector
Yeast
Zooplankton
Zoospore
Zygote

Microorganisms to Know

Gymnodinium breve
Gonyzulax species
Pfiesteria piscida
Giardia lamblia
Leishmania species
Trichomonas vaginalis
Trypanosoma brucei
Entamoeba histolytica
Balantidium coli
Plasmodium
Anopheles mosquito

Toxoplamsa gondii
Naegleria species
Cryptosporidium parvum
Coccidioides immitis
Aspergillus
Histoplasma capsulatum
Candida albicans
Saccharomyces
Penicillium
Rhizopus
Ceratocystis ulmi

Puccinia graminis
Neurospora crassa
Phytophthora infestans
Pediculus humanus
Phthirus pubis
Dermacentor andersoni
Ixodes scapularis
Demodex folliculorum
Demodex brevis
Sarcoptes scabiei
Schistosoma mansonii

Learning Activities

1.   List five characteristics of algae.

1.	
2.	
3.	
4.	
5.	

2.   Match the organism with the appropriate group.

	Organism	Group
	1. *Entamoeba*	A. Fungi
	2. *Enterobius vermicularis* (Pinworm)	B. Protozoans
	3. *Trypanosoma*	C. Algae
	4. Mosquito (*Aedes, Anopheles, Culex*)	D. Cestodes
	5. Housefly	E. Trematodes
	6. Mushroom and Puffball	F. Nematodes
	7. Tapeworm (*Taenia* sp.)	G. Arachnida
	8. *Plasmodium*	H. Insecta
	9. Yeast	
	10. Lice (*Pediculus*)	
	11. Mold	
	12. Fleas (Ex. *Xenopsylla*)	
	13. *Trichinella spiralis*	
	14. Ticks (*Dermacentor, Ixodes, Ornithodorus*)	
	15. *Gonyaulax*	
	16. *Giardia lamblia*	

3.   List two major beneficial functions of algae.

1.	
2.	

4.   Finish the table indicating the characteristics of the four groups of fungi and list an example of each group.

Group	Asexual reproduction	Sexual reproduction	Distinguishing characteristics	Example
Zygomycetes				
Basidiomycetes				
Ascomycetes				
Deuteromycetes	Not Known			

5. Finish the following table giving the disease, if any, caused by the protozoan listed.

	Protozoan	Disease
1.	*Trypanosoma*	
2.	*Giardia*	
3.	*Trichomonas*	
4.	*Leishmania*	
5.	*Entamoeba*	
6.	*Balantidium coli*	
7.	*Plasmodium*	
8.	*Toxoplasma gondii*	
9.	*Cryptosporidium*	
10.	*Microsporidium*	

6. Name the major diseases caused by the following fungi.

Fungi	Disease
Candida albicans	
Coccidioides immitis	
Filbasidiella neoformans	
Histoplasma capsulatum	
Pneumocystis carinii	
Sporothrix schenckii	

7. Explain how Deuteromycetes is different from all of the other groups of fungi.

8. What is the most common route for the transmission of helminthic diseases to humans?

_____

_____

9. For the following helminths list the disease that they cause, if any, and the characteristics of that disease.

Organism	Disease	Disease Characteristics
Enterobius vermicularis		
Trichuris trichiura		
Necator americanus		
Ancylostoma duodenale		
Strongyloides stercoralis		
Ascaris lumbricoides		
Trichinella spiralis		
Wuchereria bancrofti		
Taenia saginata or Taenia solium		
Diphyllobothrium latum		
Schistosoma mansonii		

10. You would not like to have a tapeworm for your very own. List three ways that you could prevent acquisition of a tapeworm. (Include altering the life cycle of the parasite.)

1.	
2.	
3.	

11. Identify the following arthropods as mechanical or biological vectors and name a disease with which they are associated.

Arthropod	Type of vector	Disease
Housefly		
Louse		
Flea		
Mosquito		
Tick		

Self Test

1. Which of the following statements describe algae?

    a. They are eukaryotes.
    b. They may be either microscopic or macroscopic.
    c. They are classified according to their photosynthetic pigments.
    d. All of the above statements are correct.
    e. Only a and b are correct statements.

2. Which of the following is the most likely way to get an infection of *Giardia lamblia?*

    a. injection
    b. ingestion of a cyst
    c. mosquito bite
    d. inhalation of an endospore
    e. ingestion of a trophozoite

3. Malaria is caused by which of the following organisms?

    a. *Toxoplasma*
    b. *Trypanosoma*
    c. *Trichomonas*
    d. *Giardia*
    e. *Plasmodium*

4. Malaria is transmitted to humans by

    a. ingesting cysts in water or food.
    b. inhaling trophozoites.
    c. mosquito bites.
    d. dirty fingers.
    e. handling cats.

5. A unique feature of sporozoans is that they

    a. divide by transverse binary fission.
    b. reproduce exclusively by sexual means.
    c. are not motile in the adult form.
    d. reproduce exclusively by asexual means.
    e. have two types of nuclei.

6. Which of the following genera causes vaginitis?

    a. *Toxoplasma*
    b. *Giardia*
    c. *Plasmodium*
    d. *Trichomonas*
    e. *Trypanosoma*

7. Which of the following statements describe protozoans?

    a. They are eukaryotes.
    b. They can be grouped on the basis of their means of motility.
    c. They reproduce only by asexual means.
    d. All of the above are correct statements.
    e. Only a and b are correct.

8. Which of the following statements about the Deuteromycetes are true?

    a. Their usual habitat is aquatic.
    b. Sexual reproduction is absent or unknown.
    c. Their cell walls are cellulose.
    d. All of the above statements are correct.
    e. Only a and b are correct statements.

9. Which of the following statements describe fungi?

    a. They are eukaryotes.
    b. Most fungi are anaerobic.
    c. They reproduce only by sexual means.
    d. All of the above statements are correct.
    e. Only a and b are correct statements.

10. Fungal infections of the skin are called

    a. systemic mycoses.
    b. superficial psychoses.
    c. intermediate mycoses.
    d. superficial mycoses.
    e. yeasts.

Thought Questions

1. What characteristics of algae account for their relative inability to cause disease?

2. What characteristics of fungi account for their ability to be very destructive?

3. What is the best way to control the spread of a vector-borne disease?

Answers to Self Test Questions

1-e, 2-b, 3-e, 4-c, 5-c, 6-d, 7-e, 8-b, 9-a, 10-d

Chapter 13   Viruses of Bacteria

Overview

Viruses are obligate intracellular parasites that may be associated with all forms of life. In this chapter the general architecture of viruses is presented. A variety of viral interactions with bacterial host cells are discussed with emphasis on replication of viruses. Generalized and specialized transduction are presented as well as the determinants of host range including the restriction-modification system.

Learning Objectives

After studying the material in this chapter, you should be able to:

1. Describe the composition of a virus.
2. Define:
    - Virus
    - Virion
    - Capsid
    - Nucleocapsid
    - Capsomere
    - Spikes or attachment proteins
3. Differentiate between "naked" and enveloped viruses.
4. Explain why viruses can multiply only inside cells.
5. Define:
    - Productive infection
    - Lytic phage
    - Temperate phage
    - Latent infection
    - Lysogenic state
    - Filamentous phage
6. Describe the following steps in phage replication:
    - Attachment
    - Penetration
    - Transcription
    - Replication
    - Assembly or maturation
    - Release
7. Identify:
    - Positive (+) sense strand of RNA
    - Negative (-) sense strand of RNA
8. Compare and contrast DNA phage replication with RNA phage replication.
9. Describe the integration of phage DNA into the bacterial chromosome.
10. Define or describe:
    - Immunity of lysogens
    - Lysogenic conversion

11. Explain how T4 phage prevent the infected bacteria from synthesizing bacterial proteins.
12. Describe two ways in which phage can replicate in harmony with their host.
13. Explain how filamentous phage are released from a host cell.
14. Explain how phage induction is an advantage for the phage.
15. Define transduction; differentiate between generalized transduction and specialized transduction.
16. Define defective phage.
17. Describe the host range of viruses and explain what determines it.
18. Describe the restriction-modification system.

Key Concepts

1. Viruses are acellular microbes composed of a protein coat (capsid) surrounding nucleic acid that may be either DNA or RNA.
2. Some animal viruses have a lipid bilayer (envelope) surrounding the viral particle (virion) while others viruses lack this envelope and are naked.
3. Bacteriophage are viruses that infect bacteria; they are naked in almost all cases.
4. Viruses replicate within a host cell using the host cell metabolic machinery.
5. Replication follows a pattern consisting of attachment, penetration, transcription, replication of phage DNA and proteins, assembly or maturation, and release.
6. Phage may be integrated into the host cell DNA as a prophage (lysogeny).
7. Transduction is a process by which DNA from a host cell may be conveyed by a phage to a recipient cell.
8. The host range of phage is determined by the presence of specific receptors on the host cell surface and by the restriction-modification system of the host.

Summary Outline

13.1 General **characteristics**
  A.   **Viruses** are **non-living agents** associated with all forms of life. Each **virion** consists of **nucleic acid** surrounded by a **protein coat**. They are approximately 100 to 1000 fold smaller than the cells they infect.
  B.   Virus architecture
    1.   Shapes: **Isometric, helical** or **complex**
    2.   The shape is determined by the **protein coat (capsid)** that surrounds the **nucleic acid**. These comprise the **nucleocapsid**. Each **capsid** is composed of **capsomeres; attachment proteins** or **spikes** project from the capsid.
    3.   **Enveloped viruses** have a lipid bilayer surrounding the coat.
    4.   **Naked viruses** do not have an envelope are naked.
  C.   **Viral genome**: Viruses contain either **RNA or DNA**, which may be **single-stranded or double-stranded**.
  D.   **Replication cycle**: Viruses only **multiply within living cells** using the host cell machinery. Some viruses live in harmony with their hosts and others kill the hosts.
13.2 **Virus interactions with host cells**
  A.   **Productive infections** occur when phage multiply inside bacteria and lyse the cells (virulent and lytic)
  B.   Other phage multiply but are extruded from the cell and do not kill it.

C.   **Temperate phage** integrate their DNA into the host cell where it multiplies either as a **plasmid** or integrated into the host chromosome as a **prophage** which may confer new properties on the cell.

D.   A **latent infection** may show no signs that cells are infected.

E.   **Lytic phage replication by double-stranded DNA phages**

   1.   Productive infection.

   2.   Steps:

   a)   **Attachment**—attachment proteins adsorb to specific receptors on the cell wall.

   b)   **Penetration**—the DNA is injected into the cell.

   c)   **Transcription**—the phage DNA is transcribed.

   d)   **Replication of phage DNA and proteins**

   e)   **Assembly (maturation)**—The assembly of the phage components into a complete virion.

   f)   **Release**—a phage-induced lysozyme lyses the cells releasing the phage.

   3.   **Lytic phage replication by single-stranded RNA phages**

   a)   **Phages attach** to the sex pilus.

   b)   **Replication: Viral RNA** that enters the cell acts as a **template** and synthesizes a **complementary strand**, which then serves as a **template** for the synthesis of **single-stranded RNA**.

   4.   **Phage replication in a latent state-phage lambda**

   a)   The temperate phage λ can either go through a **lytic cycle** similar to T4 or **integrate** its DNA into a specific site in the bacterial chromosome as a prophage.

   b)   **Prophage** often code for proteins, which confer unique properties on the bacteria, **lysogenic conversion**.

   5.   **Extrusion following phage replication**: **Filamentous phage** do not take over the metabolism but multiply productively as the host multiplies and are released by extrusion through the cell wall.

   6.   **Lytic infection by single-stranded DNA phage**: Single stranded DNA phages exist that can lyse cells.

13.3   **Transduction**: there are two types of transduction: **generalized** and **specialized**.

A.   **Generalized transduction**, which can be carried out by virulent and temperate phage, involves the transfer of any piece of the bacterial chromosome from one cell to another cell of the same species.

B.   **Specialized transduction** involves the transfer of specific genes.

13.4   **Host range** of phage: Factors that determine the host range of phage include **specificity of receptors** on the bacterial surfaces and the **restriction-modification system**.

Terms You Should Know

Bacteriophage
Burst size
Capsid
Capsomere
Defective phage
Enveloped viruses
Filamentous phage
Generalized transduction
Host range
Latent infection

Lysogenic conversion
Lysogenic phage
Lytic phage
Matrix protein
Naked viruses
Negative (-) sense strand of RNA
Nucleocapsid
Phage induction
Positive (+) sense strand of RNA
Productive infection

Prophage
Restriction-modification system
Specialized transduction
Spikes
Temperate phage
Virion
Virulent phage
Virus

Learning Activities

1. Define:

Term	Definition
Virus	
Virion	
Bacteriophage	
Capsid	
Nucleocapsid	
Capsomere	
Spikes	

2. Indicate which statements about viruses are true and which ones are false.

Answer	
	Viruses are extracellular parasites.
	The protein coat of a virus is called a capsid.
	Viruses are obligate intracellular parasites.
	Naked viruses lack an envelope.
	Spikes are used for adsorption or attachment.
	Spikes are important in penetration.
	The host range of viruses is very specific.
	Viruses multiply inside of host cells using the host cell machinery.
	Enveloped viruses are able to construct their own envelopes.
	Viruses can multiply only inside cells.

3. Why can viruses multiply only inside of cells?

_____

_____

4. Differentiate between positive (+) sense strand of RNA and negative (-) sense strand of RNA.

5. Put the sequence of events in viral replication of lytic double-stranded viruses in the proper order. Describe each event.

Order	Event	Description of event
	Assembly	
	Replication	
	Release	
	Penetration	
	Transcription	
	Attachment	

6. Differentiate between lysis and lysogeny.

Lysis	
Lysogeny	

7. Define or describe:

Generalized transduction	
Specialized transduction	

Self Test

1. After a bacteriophage has become integrated into the host cell chromosome it is called a(n)

  a. prophage.
  b. temperate phage.
  c. lysogen.
  d. intemperate phage.
  e. lytic phage.

2. List in the proper sequence the following stages of the lytic cycle of bacteriophages.

  1. phage DNA is synthesized
  2. lysis
  3. penetration
  4. attachment
  5. phage DNA is transcribed
  6. assembly

  a. 3, 5, 1, 2, 4, 6
  b. 4, 5, 1, 6, 2
  c. 5, 6, 2, 4, 3, 1
  d. 4, 5, 3, 1, 6, 2
  e. 4, 3, 5, 1, 6, 2

3. A complete infective viral particle is called a

   a. bacteriophage.
   b. viroid.
   c. prophage.
   d. lysogen.
   e. virion.

4. Which of the following are obligate intracellular parasites?

   a. RNA viruses
   b. DNA viruses
   c. enveloped viruses
   d. naked viruses
   e. bacteriophage
   f. All of the above.

5. A virus that has an outer double layer of lipid is called

   a. an animal virus.
   b. an enveloped virus.
   c. a naked virus.
   d. a phage.
   e. None of the above.

6. A virion is a

   a. naked, infectious piece of RNA.
   b. complete, infectious virus particle.
   c. capsid without a nucleic acid.
   d. nucleic acid without a capsid.
   e. naked, infectious piece of DNA.

7. All viruses require a host cell in order to replicate.

   a. True
   b. False

8. The transfer of genetic information by means of a bacteriophage is called

   a. transduction.
   b. transformation.
   c. transportation.
   d. transmutation.
   e. conjugation.

9. The process in which phage DNA is integrated into the host cell DNA, is called

   a. lysogeny.
   b. generalized transduction.
   c. specialized transduction.
   d. lysis.
   e. None of the above.

10. The host range of bacteriophages is determined by

   a. specific attachment proteins of the virus.
   b. specific receptors on the bacterial host.
   c. the ability of the entering phage DNA to escape degradation by host restriction-endonucleases.
   d. Both b and c are correct.
   e. a, b, and c are correct

Thought Questions

1. Describe how a virus is different from a cell.

2. Explain how phages could be used to treat a bacterial infection.

Answers to Self Test Questions

1-a, 2-e, 3-e, 4-f, 5-b, 6-b, 7-a, 8-a, 9-a, 10-d

Chapter 14   Viruses, Prions, and Viroids: Infectious Agents of Animals and Plants

Overview

While the knowledge gained through work with bacteriophages is applicable to plant and animal viruses, there are significant differences. The biology of plant and animal viruses is presented in this chapter along with methods used to study viruses. Both plant and animal viruses cause significant disease and are involved in the formation of tumors. Prions and viroids are also discussed.

Learning Objectives

After studying the material in this chapter, you should be able to:

1. Define "virus" and describe the composition of a virus, differentiate between "naked" and enveloped viruses and between phage and animal viruses.
2. List the criteria for classifying animal viruses.
3. List the groups of animal viruses based on routes of transmission.
4. Explain how plant viruses and animal viruses are cultured in the laboratory.
5. List four methods for quantifying animal viruses.
6. Describe in the proper sequence what occurs in viral replication in animal cells.
7. Describe how and where viral proteins are made.
8. Identify the event in which an enveloped virus acquires its envelope.
9. Describe how the following types of viruses are released from their host cell.
    * Naked viruses
    * Enveloped viruses
    * Phage
10. Differentiate between acute and persistent viral infections.
11. List and describe the four kinds of persistent viral infections giving specific examples of each.
12. Explain the importance of reverse transcriptase and describe what occurs in the biosynthesis of retrovirus.
13. Define:
    * Tumor or neoplasm
    * Benign
    * Malignant
    * Metastasis
    * Proto-oncogenes
    * Oncogenes
14. Describe the role of viruses in the formation of tumors.
15. List four characteristics of transformed cells in culture.
16. List several viruses that are associated with tumors in humans.
17. Describe the relationship between proto-oncogene, viral oncogene and tumor formation.

18. Explain how the host range of animal viruses may be modified by phenotypic mixing or genetic reassortment.
19. Define
    - Antigenic shift
    - Antigenic drift
20. Explain why plant viruses are of economic importance.
21. Describe how plant viruses are transmitted.
22. Define and differentiate between a viroid and a prion.
23. Describe the hosts of viroids and prions.

Key Concepts

1. Enveloped viruses have a lipid membrane acquired from the cell membrane of the host cell; naked viruses do not have this membrane.
2. Current taxonomy of animal viruses is based on genome structure, particle structure and presence or absence of an envelope.
3. Human viruses are grouped by mode of transmission into enteric, respiratory, zoonotic and sexually transmitted groups.
4. Viruses can be cultured in animals, embryonated chicken eggs and cell cultures depending on the specific virus.
5. Viruses can be quantified by plaque assay, quantal assays and hemagglutination.
6. There are two categories of viral infections: acute infections usually cause self-limited diseases in which the virus often remains localized and persistent infections in which the virus establishes infections that remain for years with or without symptoms.
7. The life cycle of an animal virus that results in an acute infection with cell lysis consists of (a) attachment, (b) penetration, (c) transcription, (d) replication of viral nucleic acid and proteins, (e) assembly and (f) release.
8. Persistent infections are divided into four groups: (a) late complications following an acute infection, (b) latent infections, (c) chronic infections and (d) slow infections.
9. Complex infections have features that are associated with more than one of the other types of infection. An example is HIV infection, which has the features of a latent, chronic and slow infection.
10. Tumors are abnormal growths of cells and can be benign or malignant; viruses can cause some types of tumors in humans.
11. Proto-oncogenes is class of host cell regulatory genes that are commonly involved in tumor formation.
12. Retroviruses transform cells by inserting transforming genes of the virus into the genome of the host cell.
13. Host range of human viruses is determined by the kinds of receptor proteins on host cell surfaces.
14. The host range can be altered if two different viruses with different host ranges infect the same cell and the genes of one virus are incorporated within the protein coat of the other during assembly.
15. Plant viruses, which may replicate in both plants and insects, may cause significant disease in plants.
16. Viroids composed of naked RNA and prions composed of proteins without nucleic acids are unconventional agents that are associated with disease in both plants and animals.

Summary Outline

14.1 Structure and Classification of Animal Viruses
    A. **Structure** of animal viruses
        1. Animal viruses consist of **nucleic acid** surrounded by a **protein coat**, the **capsid**.
        2. Some have an **envelope**, which is a **lipid bilayer** similar to the plasma membrane of the host.
        3. **Segmented viruses** contain more than one RNA molecule.
    B. **Classification** of animal viruses—Criteria
        1. **Genome structure**
        2. **Particle structure**
        3. **Presence or absence of a viral envelope**
    C. **Groupings based on routes of transmission**
        1. **Enteric viruses**
        2. **Respiratory viruses**
        3. **Zoonoses** (transferred from animals to other animals)
        4. **Sexually transmitted diseases**
14.2 **Methods used to study viruses**
    A.   **Cultivation** of host cells in **living animals**; others can be grown in **tissue culture**.
    B.   **Quantitation** by **plaque assay**, **virions counted** with the electron microscope, **quantal assays** or **hemagglutination**.
14.3 **Interactions of animal viruses with their hosts**
    A.   Most viruses do not kill their host.
    B.   **Acute infections** are **self-limited** in which the virus remains localized and diseases are of short duration and lead to lasting immunity.
    C.   The **replication cycle** is similar to that of phage T4.
    D.   The steps in the infection process include:
        1. **Attachment** to specific receptors
        2. **Entry** of the virus or **penetration** of the entire virion
        3. **Targeting** to the site of viral replication
        4. **Uncoating** – removal of the capsid
        5. **Replication** of nucleic acids and proteins
            a. **Transcription** of the viral genome into mRNA
            b. **Replication** of **virus DNA** and **proteins**
        6. **Maturation** – assembly of the virus
        7. **Release** of the virion so that it infects other cells
        8. **Shedding** – leaving the host
        9. **Transmission**
    a.   **Persistent infections** are infections in which the virions are continually present in the body and are released from cells by budding. They can be:
        1.   **Late complications** following an acute infection
        2.   **Latent infections**
        3.   **Chronic infections**
        4.   **Slow infections**
    F.   **Complex infections**: Infections with the characteristics of more than one category.
        Example: **HIV infection**, which has characteristics of a chronic, latent and slow viral infection.
14.4 **Virus-induced tumors**: General aspects
    A.   One class of host cell regulatory genes that are commonly involved in tumor formation is **proto-oncogenes**
    B.   Mutations in either of this gene class predisposes the host to tumor formation.
    C.   Proto-oncogenes are transcriptional activators

14.5 Viruses and animal tumors
- A.   **Retroviruses** are the most important tumor viruses in animals.
- B.   **Oncogenes,** which are mutants from the host's cell proto-oncogenes, can modify the properties of cells growing in tissue culture.
- C.   **Retroviruses** transform cells by inserting **oncogenes** into the genome of the host cell and interfere with the normal **intracellular control functions** of the cell's **proto-oncogenes**.

14.6. Viruses and human tumors
- A.   **Human tumors** caused by viruses are primarily caused by **double-stranded DNA tumor viruses.**
- B.   Tumor formation is promoted by the interaction of **products** of the **oncogenes** with **tumor-suppressor proteins**.
- C.   **Retroviruses** cause a rare leukemia.

14.7 Viral host range
- A.   Most viruses can infect only **certain cells** within a **single species.**
- B.   **Viruses** causing **zoonoses** can multiply in widely divergent species.
- C.   **Viruses** can **modify their host range** if two viruses with different host ranges can infect the same cell.
- D.   Protein coats may be exchanged in phenotypic mixing.
- E.   Genomes may be exchange in segmented viruses.

14.8 Plant viruses
- A.   Many **plant diseases** are caused by viruses.
- B.   **Virions enter through wound sites**.
- C.   **Spread of plant viruses** is largely by humans, by planting seeds in contaminated soils, through transfer from infected plants by grafting and through the parasitic plant, dodder, which can establish connections between an infected and uninfected plant.
- D.   **Plant viruses** may be transmitted by **insects**.

14.9 Virus-like agents
- A.   **Prions**, which consist of **protein and no nucleic acid**, have been linked to a number of **fatal neurodegenerative diseases** called transmissible spongiform encephalophathies
- B.   **Viroids**, which are plant pathogens that consist of **circular, single-stranded RNA molecules**, are about one-tenth the size of the smallest infectious viral RNA known.

Terms You Should Know

Acute infection	Exocytosis	Polycistronic mRNA
Acute infection	Fecal-oral route	Primary culture
Antigenic drift	Genetic reassortment	Prion
Antigenic shift	Hemagglutination	Protease inhibitor
Attachment proteins or spikes	Icosahderal symmetry	Proto-oncogene
Balanced pathogenicity	Latent infection	Provirus
Benign tumor	Malignant tumor	Quantal assay
Budding	Matrix protein	Retrovirus
Capsid	Metastasis	Reverse transcriptase
Capsomere	Monolayer	Segmented viruses
Carrier state	Naked viruses	Sexually transmitted viruses
Cell or tissue culture	Neoplasm	Slow infection
Chronic infection	Nucleocapsid	Transformation
Complex infection	Oncogenes	Uncoating
Cytopathic effect	Persistent infection	Viroid
Endocytosis	Phenotypic mixing	Zoonoses
Enteric viruses	Plaque assay	
Enveloped viruses	Pleomorphic	

Learning Activities

1. Indicate which statements about viruses are true and which ones are false.

Answer	
	Viruses are extracellular parasites.
	The protein coat of a virus is called a capsid.
	Viruses are obligate intracellular parasites.
	Naked viruses lack an envelope.
	Animal viruses may infect any kind of animal cell.
	Spikes are important in penetration.
	Spikes are used for adsorption or attachment.
	Spikes may cause hemagglutination.
	Viruses multiply inside of host cells using the host cell machinery.
	Enveloped viruses are able to construct their own envelopes.

2. Put in the proper order the sequence of events in viral replication of animal viruses that produce acute infections. Describe each event.

Order	Event	Description of event
	Maturation	
	Uncoating	
	Release	
	Entry (penetration)	
	Biosynthesis (biosynthesis)	
	Attachment (adsorption)	
	Target site of viral replication	
	Shedding	
	Transmission	

3. Indicate which of the following are used to culture viruses.

Used to culture viruses	Method
	Culture media
	Cell culture
	Laboratory animals
	Embryonated eggs

4. Match the virus with the appropriate family.

	Virus	Family
1.	Epstein-Barr virus and cytomegalovirus	a. *Parvoviruses*
2.	Rubella virus and viruses that cause encephalitis in humans	b. *Papovaviruses*
3.	Smallpox virus	c. *Adenoviruses*
4.	Common cold virus	d. *Herpesviruses*
5.	HIV	e. *Poxviruses*
6.	Viruses that cause warts and other tumors	f. *Hepadnaviruses*
7.	Lassa fever	g. *Picornaviruses: Enterovirus*
8.	Viruses that cause gastroenteritis after eating shellfish	h. *Picornaviruses: Rhinovirus*
9.	Poliovirus	i. *Togaviruses*
10.	Ebola and Marburg viruses	j. *Flaviruses*
11.	Gastroenteritis	k. *Orthomyxoviruses*
12.	Viruses that are associated with genital and oral carcinomas	l. *Paramyxoviruses*
13.	Rabies virus	m. *Coronaviruses*
14.	Hepatitis B virus	n. *Rhabdoviruses*
15.	Measles and mumps viruses	o. *Arenaviruses*
16.	Viruses that cause the common cold and upper respiratory tract infections	p. *Filoviruses*
17.	Hantavirus	q. *Reoviruses*
18.	Yellow fever virus	r. *Retroviruses*
19.	Norwalk agent and viruses that cause gastroenteritis	s. *Bunyaviruses*
20.	Influenza virus	t. *Caliciviruses*

5.   How are the following kinds of viruses released from host cells?

Kind of virus	Release
Naked viruses	
Enveloped viruses	

6.   Indicate which statements are true and which ones are false.

Answer	
	Herpes simplex is associated with cervical cancer.
	Varicella-Zoster virus is a herpesvirus.
	A slow viral infection is an infection in which the disease process occurs gradually over a long period of time.
	Host cell ribosomes produce viral proteins.
	Cold sores are an example of a latent viral infection.
	Creutzfeldt-Jakob disease is an example of a slow viral infection.
	Epstein-Barr virus is the causative agent of Burkitt's lymphoma, nasopharyngeal carcinoma, and mononucleosis.
	Hepadnaviridae and Retroviridae contain reverse transcriptase
	A viroid is a complete, infectious virus particle.
	A prion is an infectious piece of protein.

7.   Indicate the type of infection for each of the infections listed.

Type of infection	Infection
	Cold sores caused by herpes simplex
	Measles
	Subacute sclerosing panencephalitis
	Smallpox
	Influenza
	AIDS
	Kuru
	Cancer
	Hepatitis B
	Creutzfeldt-Jakob spongiform encephalopathy
	Common cold

Self Test

1. Penetration of host cells by animal viruses occurs by

    a. injection.
    b. exocytosis.
    c. endocytosis.
    d. spikes.
    e. a vector.

2. During which of the following stages would an envelope be acquired by an animal virus?

    a. penetration
    b. synthesis
    c. release
    d. replication
    e. assembly

3. An animal virus that has become integrated into the host cell genome is called a

    a. prophage.
    b. provirus.
    c. temperate virus.
    d. temperate phage.
    e. virion.

4. Which of the following statements about viral spikes is/are true?

    a. They contain lysozyme for rupturing the host cell.
    b. They function in recognition and attachment to host cells.
    c. They are proteins.
    d. Both b and c are correct.
    e. Both a and b are correct.

5. An RNA virus that contains reverse transcriptase, when infecting a cell will first synthesize a

    a. mRNA molecule.
    b. complementary strand of DNA.
    c. cDNA molecule.
    d. complementary strand of RNA.
    e. rRNA molecule.

6. Which of the following is a late complication of an acute infection?

    a. measles
    b. subacute sclerosing panencephalitis (SSPE)
    c. Burkitt's lymphoma
    d. Creutzfeldt-Jakob spongiform encephalopathy
    e. None of the above.

7. An example of a chronic virus infection is

    a. cancer.
    b. hepatitis C.
    c. chickenpox.
    d. herpes simplex.
    e. Two of the above are correct.

8. An example of a latent virus infection is

    a. cancer.
    b. hepatitis B.
    c. subacute sclerosing panencephalitis (SSPE).
    d. herpes simplex.
    e. Two of the above are correct.

9. An example of a slow viral infection is

    a. subacute sclerosing panencephalitis (SSPE).
    b. cancer.
    c. shingles.
    d. mad cow spongiform encephalopathy.
    e. Two of the above are correct.

10. Almost all slow viral infections involve

    a. the central nervous system.
    b. a viroid.
    c. a prion.
    d. the respiratory system.
    e. retroviruses.

11. Humans are major vectors for plant viruses.

    a. true
    b. false

12. Which of the following statement a true for plant viruses?

    a. Plant viruses cause many plant diseases.
    b. Plant viruses invade through wound sites in the cell wall.
    c. Infections by plant viruses usually are recognized by outward signs.
    d. Only a and b are correct.
    e. a, b, and c are correct.

Thought Questions

1. Explain the criteria used for the classification of animal viruses.

2. Justify or refute the statement that viruses cause cancer.

3. Compare and contrast normal and malignant cells.

Answers to Self Test Questions

1-c, 2-c, 3-b, 4-d, 5-d, 6-b, 7-b, 8-d, 9-d, 10-a, 11-a, 12-e

Chapter 15   The Innate Immune Response

Overview

Host defense mechanisms are of two types. Innate immune response mechanisms are nonspecific and general in function. These mechanisms defend the host against any invading microorganism. Adaptive immune mechanisms are specific and defend the host against microbes to which the body has been specifically sensitized. In this chapter the innate immune mechanisms such as physical barriers, complement system, phagocytosis, inflammation, interferons, and fever are presented.

Learning Objectives

After studying the material in this chapter, you should be able to:

1. Differentiate innate immunity and adaptive immune response.
2. Define or identify
   - Self
   - Lysozyme
   - Perosidase
   - Lactoferrin
   - Surface receptors
   - Cytokines
   - Lymphokines
   - Colony-stimulating factors (CSFs)
   - Interferons (IFNs)
   - Interleukins (Ils)
   - Tumor necrosis factors (TNFs)
3. Define first-line defense.
4. List and describe the mechanical and chemical factors that are involved in preventing the invasion of microorganisms through the skin and mucous membranes.
5. Describe the role of normal flora in innate immunity.
6. Identify and describe of the roles of the following kinds of cells:
   - Neutrophils
   - Basophils
   - Eosinophils
   - Monocytes
   - Macrophages
   - Lymphocytes
7. Define complement.
8. Differentiate among the classical, lectin, and alternate complement pathways.
9. List the outcomes of complement fixation.
10. Define and describe the process of phagocytosis.
11. Define
    - Chemotaxis
    - Phagolysosome
    - Opsonization

12. Define inflammation; give its functions and cardinal signs.
13. Describe the effects of the pro-inflammatory cytokines
14. Define or identify:
    - Diapedesis
    - Acute inflammation
    - Chronic inflammation
    - Apoptosis
15. Identify interferon and explain what it does.
16. Define fever and explain how it is induced.

Key Concepts

1. Innate immunity is the body's first line of defense against foreign invaders.
2. The skin and mucous membranes form important barriers to microbes.
3. Secretions including mucus, saliva, perspiration, gastric juice and urine act physically and chemically as a barrier to microbes.
4. Normal flora help to prevent infections by competing with potential pathogens.
5. The complement cascade is a set of proteins that when activated by antigen-antibody complexes or by certain bacterial cell wall components cause inflammation, opsonization and cell lysis.
6. Cytokines have roles in the inflammatory response, contribute to fever induction, and are involved in the initiation of specific immune responses.
7. Interferons are glycoproteins that function against viruses.
8. Phagocytosis is a process in which foreign materials including microbes are engulfed and destroyed by specialized host cells.
9. Inflammation is a nonspecific response to tissue damage that functions to deliver more defense materials and cells to damaged tissue and to contain the infection thereby setting up conditions for repair.
10. Fever is an abnormally high body temperature induced by pyrogens, which decreases the growth rate of some bacteria and increases the production of host defense cells and substances.

Summary Outline

15.1 **Innate Defenses**
   A. Composed of first-lined defenses, toll-like receptors, complement and phagocytes.
   B. **Inflammation** is a coordinated response that involves innate defenses.
15.2 **First-line Defenses**
   A. **Physical barriers** include the skin and mucous membranes
   B. **Antimicrobial substances** such as **lysozymes, peroxidase enzymes, lactoferrin,** and **defensins** inhibit or kill microorganisms.
   C. **Normal Flora** competes with pathogens for the same niche and stimulates the host defenses.
15.3 Cells involved in host defense
   A. The granulocytes are **eosinophils**, **basophils** and **neutrophils** (PMNs).
   B. The **mononuclear phagocytes**, **monocytes** and **macrophages**, along with the **neutrophils** are **phagocytic**.
   C. **Macrophages** and **dendritic cells** also play essential roles in specific immune responses.
   D. **Lymphocytes**
      1. **Natural killer cells** are lymphocytes that kill abnormal cells in a non-antigen-specific manner.

2.   **T cells** are responsible for **specific cellular immune responses**.

3.   **B cells** are responsible for producing **antibodies**.

15.4 Cell communication

A. **Surface receptors** bind ligands on the cell surface and thus signal the cell of the presence of the ligands.

B. **Cytokines**

1.   **Cytokines** are **small regulatory proteins** essential for communication between cells. They have many actions in the **inflammatory response** and they contribute to **fever production** and to **development of specific immune responses**.

2.   **Cytokines** include **interleukins (Ils), colony-stimulating factors (CSFs), tumor necrosis factors (TNFs), chemokines,** and **inferferons.**

3.   **Interferons** are **antiviral glycoproteins** important in activating macrophages and in development and regulation of specific immune responses.

4.   **Interleukins** are cytokines produced by leukocytes that function in the induction of fever, signaling release of PMNs, attracting leukocytes into areas of inflammation and inducing proliferation of lymphocytes.

5.   **Colony stimulating factors** direct immature cells into appropriate maturation pathways.

6.   **Tumor necrosis factors** induce fever, recruit neutrophils into areas of inflammation, and are antiviral; some can kill target cells.

C. **Adhesion molecules** are chemicals that allow cells to adhere to other cells.

15.5 **Sensor systems**

A. **Toll-like receptors** enable cells to detect molecules signaling an invading microorganism.

B. The **complement system**

1.   Complement is a **group of blood proteins** that act in a cascading fashion, leading to an amplified effect.

2.   Complement is activated by a specific antigen-antibody reaction in the classical pathway, or nonspecifically by bacterial products in the alternative pathway and the lectin pathway.

3.   The end of complement activation may result in **inflammation, opsonization** and **cell lysis**.

15.6 **Phagocytosis**

A. Process of **phagocytosis** occurs in several steps:

1.   **Chemotaxis**

2.   **Recognition** and **attachment**

3.   **Engulfment**

4.   **Fusion** of the **phagosome** with the **lysosome**

5.   **Destruction** and **digestion**

6.   **Exocytosis**

B. **Macrophages**

1.   **Macrophages** are always present in tissues, but can signal for the migration of other macrophages as needed.

2.   **Macrophages** can be activated which increases their ability to kill.

3.   **Macrophages, giant cells,** and **T-helper** cells form groups called **granulomas** that wall off microorganism and other material not destroyed by macrophages.

C. **Neutrophils** are the first cell type to arrive at an area of damage.

15.7 **Inflammation** is an innate tissue response to any injury, characterized by swelling, heat, redness, and pain.

1.   It is an attempt by the body to contain a site of damage, localize the response, and restore tissues.

2.   Upon injury, **cytokines** and other pro-inflammatory mediators are released that lead to

a. **Dilation of local small blood vessels**

b. **Chemotaxis of leukocytes**

c. **Leakage of fluids into the tissues**
d. **Migration of leukocytes from the bloodstream into the tissues**

3. **Acute inflammation** is local, begins immediately upon injury, and increases in intensity over a short time.
4. **Chronic inflammation** involves the formation of granulomas and may last for years.
5. **Apoptosis** is programmed cell death of "self" cells that by-passes the inflammatory response

15.8 **Interferons** induce cells in the vicinity of a virally infected cell to block protein synthesis if they become infected with a virus.

15.9 **Fever** is caused by pro-inflammatory cytokine known pyrogens that are released by macrophages and act on the hypothalamus to increase body temperature. Fever decreases the growth rate of many bacteria and increases the rate of production of host defense cells and substances.

Terms You Should Know

Alternative pathway
Antibody
Antigen
Apoptosis
B cell
Basophil
Chemokine
Chemotaxis
Classical pathway
Colony-stimulating factors (CSF)
Complement
Cytokines
Defensin
Dendritic cell
Diapedesis
Endogenous pyrogens
Engulfment

Eosinophil
Exogenous pyrogens
Fever
Granulocyte
Granuloma
Hematopoietic stem cell
Histamine
Immunology
Inflammation
Innate immune response
Interferons
Interleukins
Lactoferrin
Lectin pathway
Leukocyte
Ligand
Lymphocyte
Lysosome

Lysozyme
Macrophage
Mast cell
Monocyte
Natural killer (NK) cell
Neutrophil
Normal flora
Opsonization
Peroxidase
Phagocyte
Phagocytosis
Phagolysome
Phagosome
T cell
Toll-like receptors
Tumor necrosis factors (TNF)

Learning Activities

1. Define the following terms:

Term	Definition
Innate immunity	
First-line defenses	
Complement	

2. List the mechanical and chemical factors involved in preventing the invasion of microorganisms through the skin and mucous membranes.

Mechanical factors	Chemical factors

3. List the kinds of cells that are phagocytic or become macrophages.

1.	
2.	
3.	

4. Indicate whether each of the following is involved with innate immunity, adaptive immunity, or both.

Answer	
	Inflammation
	Antibodies
	Antigens
	Memory response
	Eosinophils
	Basophils
	Mast cells
	Neutrophils
	Macrophages
	Natural killer (NK) cells
	B cells
	T cells
	Complement
	Interferon
	Fever

5. Describe the functions of the following kinds of cells.

Cell	Function
Eosinophils	
Basophils	
Mast cells	
Neutrophils	
Macrophages	
Natural killer (NK) cells	

6. Describe the differences among the classical, alternative and lectin pathways of complement activation.

_____

_____

7. List the outcomes of complement fixation.

8. List the function of the following cytokines.

Cytokine	Function
Alpha interferon	
Interleukin-1	
Interkeukin-2	
Interkeukin-4	
Interkeukin-6	
Colony-stimulating factors (CSFs)	
Alpha tumor necrosis factor (TNFs)	

9. List and describe each step of phagocytosis.

Steps of phagocytosis	Description

10. List the cardinal signs of inflammation and explain the source of each one.

Cardinal signs	Explanation

11. Describe the purpose of inflammation.

12. Complete the table regarding fever.

Definition	
Induction	
Result	

Self Test

1. Adaptive immune responses are mostly highly developed in

   a. mammals.
   b. birds.
   c. reptiles.
   d. Both a and b are correct.
   e. Both a and c are correct.

2. All blood cells in the body originate from

   a. spleen cells.
   b. liver cells.
   c. cells of the lymphatic system.
   d. hematopoietic stem cells.
   e. None of the above.

3. Which of the following blood cells typically increase in numbers during an active infection?

   a. neutrophils
   b. basophils
   c. eosinophils
   d. monocytes
   e. lymphocytes

4. Which of the following blood cells are the first to proliferate in response to an infection?

   a. eosinophils
   b. basophils
   c. neutrophils
   d. monocytes
   e. lymphocytes

5. Which of the following substances induce fever?

   a. interferon
   b. glucose
   c. pyrogens
   d. C3
   e. All of the above.

6. Which of the following are innate means of defense against invasion by microbes?

   a. intact skin
   b. flushing action of the urinary tract
   c. inflammation
   d. fever
   e. All of the above are nonspecific means of defense.

7. Which of the following substances is a normal component of serum?

   a. complement
   b. lysozyme
   c. interferon
   d. collagenase
   e. lysosome

8. Complement is activated by

   a. a reaction between antigens and antibodies.
   b. an endotoxin or microbial polysaccharide.
   c. lectin binding with mannose in the bacterial cell surface.
   d. Only a and b are correct.
   e. a, b, and c are correct.

9. After invading bacteria are engulfed by phagocytes, they are contained within a vacuole called a

   a. lysosome.
   b. phagosome.
   c. phagolysosome.
   d. lysozyme.
   e. pinosome.

10. Fever

    a. is caused by pyrogens.
    b. decreases the growth rate of many bacteria.
    c. increases the rate of production of host defense cells and substances.
    d. affects the availability of iron to microorganisms.
    e. Only a, b and c are true.

Thought Questions

1. Describe the inflammation process and explain how it defends the host against pathogens.

2. Some diseases such as rheumatoid arthritis are characterized by inflammation. How can inflammation be part of host defense and also a characteristic of disease?

Answers to Self Test Questions

1-d, 2-d, 3-a, 4-c, 5-c, 6-e, 7-a, 8-e, 9-b, 10-e

Chapter 16   The Adaptive Immune Response

Overview

The adaptive immune response is characterized by both specificity and memory. There are two types of specific immunity—humoral or B cell immunity and cell-mediated or T cell immunity. These two types do not function independently of each other, but work together. In this chapter the mechanisms of adaptive immunity are presented as is currently understood.

Learning Objectives

After studying the material in this chapter, you should be able to:

1. Identify the following types of cells:
   - B lymphocytes
   - Plasma cells
   - Memory cells
   - T lymphocytes
   - T-cytotoxic cells
   - T-helper cells
   - Effector T-cells
2. Define:
   - Antigen
   - Immunogen
   - Epitopes (antigenic determinants)
   - Antibody
   - Humoral response
   - Cellular immune response
   - Antigen-presenting cells (APC)
3. Differentiate between the primary response and the anamnestic response.
4. Differentiate between humoral immunity and cellular immunity.
5. Describe the general structure of an antibody molecule and the functions of specific regions of the molecule.
6. List and describe seven outcomes of antibody-antigen binding.
7. List the classes of immunoglobulins (antibodies), where each is produced, and its function.
8. Differentiate between clonal selection and colonal expansion.
9. Differentiate between T-dependent antigens and T-independent antigens.
10. Explain how maternal antibodies protect a baby and how long this protection lasts.
11. Explain why the human immune system can respond to millions of different antigen epitopes.
12. Define or identify
    - Major histocompatibility complex
    - MHC Class II molecules
    - Antigen presentation

- Affinity maturation
- Class switching
- Formation of memory cells
- CD markers (molecules)
- Antigen presenting cells

13. Differentiate CD4 and CD8 cells and describe their functions.
14. Describe the activation of T cells.
15. Describe the interaction among cells of the immune system in the:
    - Humoral response
    - Cellular immune response
16. Identify NK (natural killer) cells and explain how they function.
17. Explain how the immune system is capable of producing a nearly infinite antibodies and antigen-specific receptors.

Key Concepts

1. Adaptive immunity is the type of immunity that confers protection against one particular kind of microorganism.
2. Immunity to one disease does not normally confer immunity to another disease.
3. Adaptive immune responses are highly antigen-specific, have memory, and involve lymphocytes acting together with macrophages and dendritic cells.
4. Adaptive immunity depends upon two types of lymphocytes, B cells that produce an antibody response, and T cells that produce a cellular immune response. These cells interact with each other and with other cells.
5. Adaptive immunity has memory, which enables the system to act quickly against foreign material such as microbes that it has encountered previously.
6. The adaptive immune response can normally distinguish between self and nonself.
7. Antigens are large molecules that contain many surface epitopes, are usually foreign to the host, and react specifically with antibodies or immune cells.
8. Antibodies are immunoglobulins that react specifically with the antigen that induced their formation. There are five classes: IgA, IgD, IgE, IgG, and IgM.
9. Lymphocytes are differentiated by cell surface molecules. B lymphocytes mature into plasma cells, the principal antibody producers. T cells include helper T cells (CD4 cells), cytotoxic T cells (CD8 cells), suppressor T cells, and memory T cells.
10. MHC (major histocompatibility proteins) class II proteins are located on cell surfaces of antigen-presenting cells, dendritic cells, and B cells.
11. Specific antibodies on B cell surfaces recognize a single epitope to induce an antibody response. The two kinds of antigens are (a) T-cell-dependent antigens that require CD4 T helper cells to elicit an antibody response and (b) T-cell-independent antigens that induce antibody formation without help from T cells.
12. The clonal selection theory states that an antigen selects the lymphocytes specific for that antigen resulting in the proliferation of antigen-specific effector and memory cells.
13. T cytotoxic (Tc) cells respond to protein antigens on intact cells, complexed with MHC class I antigen. Tc cells destroy virus-infected and cancer cells by releasing the contents of the Tc cell granules. Cytokines that are produced by CD4 (Th1) activate macrophages.

Summary Outline

16.1 Strategy of the adaptive immune response
- A. **Adaptive immunity** is acquired throughout life.
- B. **Humoral immunity** is mediated by **B-lymphocytes.**
  1. B-lymphocytes are activated in response to extracelluar antigens.
  2. B-lymphocytes proliferate and differentiate into **plasma cells** that produce **antibodies.**
- C. **Cellular immunity** is mediated by **T-lymphocytes.**
  1. **T-cytotoxic cells** destroy host cells that harbor intracellular agents such as viruses by inducing **apoptosis.**
  2. **T-helper cells** potential cellular and humoral responses.

16.2 Anatomy of the lymphoid system
- A. Lymphatic vessels - Lymph, which may contain antigens that have entered tissues, flows in the lymphatic vessels to the lymph nodes.
- B. Secondary lymphoid organs – Locations where lymphocytes gather to contact antigens.

16.3 The nature of **antigens**
- A. **Antigens** are large, usually **foreign, molecules** with surface **epitopes** that **react specifically** with **antibodies and immune cells**.
- B. Most antigens are immunogens that can induce the production of specific antibodies or immune cells.

16.4 The nature of **antibodies**
- A. **Antibodies** are **proteins (immunoglobulins)** with **two heavy** and **two light polypeptide chains**, which **react specifically** with the antigen that induced their formation.
- B. Properties of antibodies
  1. **Antibody monomers** have a Y shape with an **antigen-binding (Fab) site** at the end of each arm of the Y and an **Fc region**, which accounts for many of the biological functions of the antibody, unique to each class.
  2. **Noncovalent, short-range bonds** hold antigen and antibody together. The reaction is **reversible**.
- C. Outcomes of **antigen-antibody binding**
  1. **Neutralization**
  2. **Immobilization** and **prevention of adherence**
  3. **Agglutination** and **precipitation**
  4. **Opsonization**
  5. **Complement activation**
  6. **Antibody-dependent cytoxicity.**
- D. **Immunoglobulin classes**
  1. **IgG** is a **monomer** that can cause **opsonization, agglutination, precipitation, complement fixation, ADCC** and **neutralization of toxins and viruses**. It is the only class of immunoglobulins that **can cross the placenta**.
  2. **IgM**, usually a **pentamer**, is the **first** class of immunoglobulins **produced during an immune response**. It is very efficient in **agglutination, precipitation, opsonization** and **complement activation**.
  3. **IgA** is abundant as a **dimer in secretions**. It **inhibits adherence of organisms to host cells,** and **protects mucous membrane surfaces**.
  4. **IgD** is a **monomer found on B cell surfaces** that acts as a **receptor** for the specific antigen it recognizes.
  5. **IgE** is a **monomer** that binds strongly to mast cells and basophils and helps to **protect against some multicellular parasites and contributes to many allergic reactions**.

16.5   **Clonal selection** of lymphocytes
  A. Antigens select lymphocytes specific for that antigen, resulting in **proliferation** of expanded clones of **antigen-specific effectors and memory cells**.
  B. Lymphocytes recognize antigens based on the antigen receptors on their cell surfaces.
  C. Lymphocytes may be **immature, naïve, activated, effector,** or **memory cells.**

16.6   **B-lymphocytes** and the **antibody response**
  A. Response to **T-dependent antigens**
    1. When T-dependent antigens bind with B-cell receptors, the antigen is internalized, degraded into peptide fragments and presented to **helper T-cells (antigen presentation)**.
    2. The help T-cell recognizes the antigen and deliver cytokines to the B cell initiating **clonal expansion** that results in the production of **plasma cells** that produce specific **antibodies** to the antigen.
    3. Under the direction of T-helper cells, the expanding B-cell population will undergo affinity maturation and class switching, and formation of **memory cells**.
    4. **Primary response** – A lag period occurs before antibodies can be detected.
    5. **Secondary response** – Memory cells are responsible for a much faster and effective response resulting in the elimination of invaders before they have the opportunity to do much harm.
  B. Response to **T-independent antigens**
    1. T-independent antigens include polysaccharides that have multiple identical evenly space epitopes and LPS.
    2. T- independent antigens can induce antibody formation without help from T cells.

16.7   **T lymphocytes**: Antigen recognition and response
  A. T-cell receptors recognize antigens presented by the **major histocompatibility (MHC) molecules.**
  B. Functions of effector **T-cytotoxic (CD8) cells**
    1. **T-cytotoxic cells** induce **apoptosis** in cells infected with a virus or other intracellular microorganism or cancerous cells.
    2. **T-cytotoxic cells** produce cytokines that cause neighboring cells to become more active against intracellular invaders.
  C. Functions of effector **T-helper (CD4) cells**
    1. **T-helper cells** respond to exogenous antigens that are presented by MHC class II molecules.
    2. **Th1 cells** judge antigens presented by macrophages.
    3. A responding **Th1 cell** activated the presenting macrophage and secretes cytokines that help direct the immune response.
    4. **Th2 cells** judge antigens presented by B cells.
    5. A responding **Th2 cell** activated the B cell and supports proliferation and most types of class switching by activated B cells.
  D. **Activation of T cells**
    1. **Native T cells** require supporting signals for activation.
    2. Activated T cells stimulate their own proliferation and becomes effective.
    3. **Dendritic cells** sample material in tissues and journey to lymphoid organs to present the antigens to naïve T cells.
    4. **Dendritic cells** that detect molecules that indicate an invading microorganism produce co-stimulatory molecules and activate both subsets of T cells.
    5. **Activated macrophages** that have engulfed foreign antigens produce co-stimulatory molecules to activate T-helper cells.

16.8 **Natural killer (NK) cells**
   A. **NK cells** mediate **antibody-dependent cellular cytotoxicity (ADCC)**.
   B. **NK cells** will kill any cells that do not have **MHC class I molecules** on their cell surfaces
16.9 **Lymphocyte development**
   A. Generation of diversity – Mechanisms include:
       1. **Rearrangement of gene segments**
       2. **Imprecise joining of gene segments**
       3. **Combinations of heavy and light chains**
   B. **Negative selection** of self-reactive B cells
       1. Negative selection occurs while B cells develop in the bone marrow.
       2. Negative selection is the process of eliminating B cells that recognize "self" molecules.
   C. **Positive and negative selection** of self-reactive cells
       1. Positive selection allows only those T cells that recognize the MHC molecules to develop.
       2. Negative selection results in the elimination of any native T cell that recognizes antigens presented by an antigen-presenting cell that does not have co-stimulatory molecules.

Terms You Should Know

Adaptive acquired immune response
Antibody
Antigen
Antigen-binding site
Antigen-presenting cell
Autoantibodies
B cell
CD markers
CD4 T cells
CD8 T cells
Cellular immunity
Clonal expansion

Clonal selection
Constant region
Effector T cell
Epitope (antigenic determinant)
Fab region
Fc region
Hapten
Humoral immunity
Immunoglobulins
Major histocompatibility complex (MHC)
Memory
Memory cell

MHC Class II
Plasma cell
Primary response
Secondary (anamnestic) response
T cell
T-Cytotoxic cell
T-dependent antigens
Th1 cell
Th2 cell
T-independent antigens
Titer cytotoxins
Tolerance T-Helper cell
Variable region

Learning Activities

1. Draw a typical antibody and label the specific regions.

2.   Define:

Term	Definition
Antigen	
Epitopes (antigenic determinants)	
Hapten	
Antibody	
Primary response	
Anamnestic response	
Antigen-presenting cell (APC)	
Clonal selection	
Humoral response	
Cellular immune response	

3.   Differentiate between:

Humoral immunity	
Cell-mediated immunity	

4.   Write the function of the following specific regions of an antibody.

Fab region	
Fc or stem region	

5.   For the following classes of immunoglobulins (antibodies) their functions and the location where they are active.

Class	Active location	Functions
IgG		
IgM		
IgA		
IgD		
IgE		

6. Explain how maternal antibodies protect a baby and how long this protection lasts.

_____

_____

7. Describe the functions of the cells involved in humoral immunity.

Cells	Functions
B cell	
Plasma cell	
Memory B cell	
T-Helper cell	

8. Describe the specific type of antibodies that are produced when one is exposed to an antigen.

First Exposure	
Second Exposure	

9. Describe the functions of the cells involved in cell-mediated immunity.

Cell	Function
T-Cytotoxic cell	
T-Helper cell	
Memory T cell	
Macrophage	

10. Describe the outcomes of innate and adaptive immune mechanisms.

Innate immunity	
Adaptive immunity	

Self Test

1. Which of the following types of chemicals would be the best antigens?

   a. lipids
   b. glucose
   c. fatty acids
   d. proteins
   e. sugars

2. Which part of the antibody molecule reacts with the antigenic determinants of the antigen?

   a. hinge
   b. heavy chains
   c. constant regions of the heavy and light chains
   d. light chains
   e. variable regions of the heavy and light chains

3. An antigen can be best defined as

   a. a hapten that combines with an antibody.
   b. a substance that elicits an antibody response and can combine specifically with these antibodies.
   c. a small molecule that attaches to cells.
   d. a carbohydrate.
   e. a protein that combines with antibodies.

4. Which of the following statements is true for B cells?

   a. They differentiate into antibody-producing cells.
   b. They function in cell-mediated immunity.
   c. They differentiate into neutrophils.
   d. They differentiate into macrophages.
   e. They function in native immunity.

5. Which of the following statements is true for T cells?

   a. They differentiate into antibody-producing cells.
   b. They function in cell-mediated immunity.
   c. They differentiate into neutrophils.
   d. They differentiate into macrophages.
   e. They function in native immunity.

6. The rapid increase and sustained production of antibodies in response to a second or subsequent exposure to an antigen is known as

   a. a primary response.
   b. the killer response.
   c. a memory or anamnestic response.
   d. inflammation.
   e. a dose response.

7. An immunoglobulin is a(n)

   a. histamine.
   b. antigen.
   c. macrophage.
   d. antibody.
   e. carbohydrate.

8. The portion of an antigen molecule that stimulates an antibody response is called the

   a. antigenic determinant.
   b. epitope.
   c. antibody determinant.
   d. Both a and b are correct.
   e. Both b and c are correct.

9. Which of the following classes of immunoglobulins can cross the placenta?

   a. IgG
   b. IgM
   c. IgA
   d. IgD
   e. IgE

10. Which of the following cell types process and present antigens to lymphocytes?

    a. B cells
    b. macrophages
    c. T cells
    d. helper T cells
    e. natural killer cells

11. Which of the following classes of immunoglobulins are secretory in function?

    a. IgG
    b. IgM
    c. IgA
    d. IgD
    e. IgE

12. Plasma cells are

    a. mature T cells.
    b. immature macrophages.
    c. antibody-producing cells.
    d. mature macrophages.
    e. immature T cells.

13. B lymphocytes originate in

    a. adult bone marrow.
    b. thymus.
    c. fetal liver.
    d. Both a and b are correct.
    e. Both a and c are correct.

14. Antibody molecules are

    a. carbohydrates.
    b. lipids.
    c. proteins.
    d. nucleic acids.
    e. None of the above.

Thought Questions

1. Describe how the adaptive immune response is fundamentally different from innate defense mechanisms.

2. Explain the function of a macrophage and how it interacts with other cells of the immune system.

Answers to Self Test Questions

1-d, 2-e, 3-b, 4-a, 5-b, 6-c, 7-d, 8-d, 9-a, 10-b, 11-c, 12-c, 13-e, 14-c

Chapter 17   Applications of Immune Responses

Overview

In this chapter the practical applications of immune responses are covered. The principles of immunization, the use of vaccines and the development of new vaccines are discussed first. Then the principles of immunological testing and the use of antigen-antibody reactions and lymphocyte response in the diagnosis of disease through serological and cellular tests are presented.

Learning Objectives

After studying the material in this chapter, you should be able to:

1. Define
   - Immunology
   - Serology
   - Monoclonal antibodies
   - Hybridomas
   - Seroconversion
   - Titer
2. Distinguish between active and passive immunity.
3. Define
   - Immune serum globulin (gamma globulin)
   - Vaccine
   - Adjuvants
   - Attenuated agents
   - Herd immunity
   - Toxoid
4. List and describe the major types of vaccines and give specific examples of each type.
5. Compare and contrast the Salk and Sabin polio vaccines.
6. Describe the following types of serological reactions and list specific applications of each:
   - Immunodiffusion
   - Immunoelectrophoresis
   - Direct agglutination
   - Indirect agglutination
   - Hemagglutination inhibition
   - Direct fluorescent antibody test
   - Indirect fluorescent antibody test
   - Complement fixation
   - Neutralization tests

7. Describe the following techniques and give examples of uses for each:
   - Radioimmunoassay (RIA)
   - Enzyme-linked immunosorbent assay (ELISA)
   - Western blot
8. Describe the following tests used in cellular immunology:
   - Identification of subsets of lymphocytes
   - Lymphocyte response to mitogens
   - Cytotoxic T cell function
   - Cell-mediated immunity to infectious agents

Key Concepts

1. Levels of protection can be determined by titer determination that is based on serial dilutions.
2. Immunization can be either passive, giving an individual antibodies, or active, causing an individual to produce antibodies.
3. Injection of antibodies gives immediate but short-lived protection.
4. Injection of immunizing agents produces longer-lasting immunity and also involves both the humoral response and the cell-mediated response.
5. Immunizing agents are found in vaccines and may include killed or attenuated microbes or subunits, which are portions of the microbial agent.
6. Recombinant DNA technology has made it possible to produce more effective and safer vaccines.
7. The specificity of immunological reactions is the basis of many diagnostic tests.
8. Serology is the use of serum antibodies to detect and measure antigens or the use of antigens to detect serum antibodies.
9. Monoclonal antibodies can be produced for immunological tests, treatment, and research, by production of hybridomas.
10. Hybridomas are clones of cells resulting from the fusion of normal antibody-producing B lymphocytes with myeloma tumor cells.

Summary Outline

17.1 Principles of immunization
   A. **Immunity** is either **natural** or **artificial**, **passive** or **active**.
   B. **Passive immunity** occurs naturally from mother to fetus and artificially by transfer of preformed antibodies, as in hyperimmune globulin.
   C. **Active immunity** occurs naturally in response to infections or other natural exposure to antigens and artificially in response to vaccine administration.
17.2 Vaccines and immunization procedures
   A. A **vaccine** is a preparation of living or inactivated microorganisms or viruses or their components used to **induce active immunity**.
   B. **Attenuated immunizing agents** are antigenic and can replicate, but are modified to be incapable of causing disease under normal circumstances.
   C. **Inactivated vaccines** may contain inactivated whole agents or subunits of the agent.
   D. **Recombinant vaccines** are genetically engineered.
   E. **Adjuvants** increase the intensity of the immune response to the antigen in a vaccine.

F. Routine childhood immunizations have prevented millions of cases of disease and many deaths during the past decades. Universal immunization is essential to eradicate some diseases and to preserve herd immunity against others.

17.3   Principles of immunological testing

A. **Serology** uses antibodies, usually in serum or other body fluids, to detect and identify antigens, or conversely, uses known antigens to detect antibodies.

B. **Seroconversion** is the change from negative to positive from specific antibodies during an infection.

C. A rise in **titer** is characteristic of an active infection.

D. **Serial dilution** of specimens permits **quantification of antibodies** in the sample.

17.4   **Precipitation reactions** occur when **soluble antigens** interact with **antibodies** in optimal proportions to cause cross-linking into a large **insoluble lattice.**

A. **Immunodiffusion tests** are precipitation tests done in gels.

B. **Immunoelectrophoresis** is a test in which mixtures are separated by electrophoresis before adding antibodies to identify the separated antigens.

17.5   **Agglutination reactions** depend on cross-linking of particulate antigen by antibody molecules to form readily visible clumps.

A. **Direct agglutination tests: Particulate antigen** reacts directly with antibodies.

B. **Indirect agglutination tests: Soluble antigen** is coated onto **particles** to give indirect agglutination.

C. **Hemagglutination inhibition: Antibodies interfere with viral agglutination of red blood cells**.

17.6   **Immunofluorescense tests**: Fluorescent dyes are used to visualize antibodies under the fluorescence microscope

A. **Direct fluorescent antibody test: Antibodies tagged** with fluorescent dyes react directly with antigen.

B. **Indirect fluorescent antibody test: Antigen** and **antibody** interact and the **complex** is detected with **fluorescent-labeled antibodies** against the immunoglobulin in the complex.

17.7   **Radioimmunoassay (RIA), Enzyme-linked immunosorbent assay (ELISA) and Western blot**

A. **Radioimmunoassay (RIA)** is based on competition for specific antibody in a test sample between known amounts of radioactively labeled antigen and unknown amounts of unlabeled antigen.

B. **Enzyme-linked immunosorbent assay (ELISA):** Enzymes that give a color reaction are used as labels in the ELISA test.

C. **Western blot** combines electrophoresis with ELISA to separate and identify protein antigens in a mixture.

17.8   The **complement fixation test** and **neutralization tests**

17.9   **Tests used in cellular immunology**

A. **Fluorescent-labeled monoclonal antibodies** are used to identify subsets of lymphocytes by microscopy or by separation in a cell sorter.

B. **Lymphocyte response to mitogens** can distinguish subsets of lymphocytes.

C. **Cytotoxic T cell function**

D. **Cell-mediated immunity to infectious agents**: Lymphocyte proliferation in response to specific antigens is measured by incorporation of radioactive thymidine into DNA.

Terms You Should Know

Active immunity
Adjuvant
Attenuated vaccine
Conjugate vaccine
DNA-based vaccine
Edible vaccine
Fluorescence-activated cell
    sorter (FACS)
Herd immunity

Hybridoma
Hyperimmune globulin
Immune complex
Immune serum globulin
Immunization
Immunoassay
Inactivated vaccine
Monoclonal antibodies
Passive immunity

Recombinant vaccine
Serial dilution
Seroconversion
Serology
Subunit vaccine
Titer
Toxoid
Vaccine
Whole agent vaccine

Tests to Know

Agglutination reactions
    Direct
    Indirect
Cellular immunology tests
Complement fixation test

Hemagglutination inhibition
Immunodiffusion tests
Immunoelectrophoresis
Immunosorbent assay (ELISA)
Neutralization tests

Immunofluorescence test
    Direct
    Indirect
Radioimmunoassay (RIA)
Western blot

Learning Activities

1. Explain how one could acquire the following types of immunity.

	Active	Passive
Natural		
Artificial		

2. Define the following:

Term	Definition
Immune serum globulin (gamma globulin)	
Vaccine	
Adjuvant	
Attenuated agent	
Herd immunity	
Toxoid	

3.   Describe the major types of vaccines.

Type of vaccine	Description
Attenuated whole-agent	
Inactivated whole-agent	
Subunit	
Recombinant	
Acellular	
Conjugated	
Nucleic-acid	

4.   Match each vaccine with its type of vaccine.

	Vaccine	Type
	1. Influenza	A. Attenuated virus
	2. Hepatitis B	B. Killed virus
	3. Chickenpox	C. Recombinant
	4. BCG	D. Acellular
	5. Pertussis	E. Conjugated
	6. Tetanus	F. Toxoid
	7. Hib (Hemophilus influenzae type b)	G. Attenuated bacterial
	8. Diphtheria	
	9. Salk polio	
	10. Sabine polio (OPV)	
	11. Rabies	
	12. MMR	

5. Differentiate between the Salk and Sabine vaccines.

	Salk	**Sabine**
Administration route		
Advantages		
Disadvantages		

6. Define the following:

Term	**Definition**
Serology	
Monoclonal antibodies	
Hybridomas	
Seroconversion	
Titer	
Immunoassay	

7. Describe the following kinds of serological reactions.

Type of reaction	**Description**
Precipitation reaction	
Agglutination reaction	
Immunofluorescence	
Complement fixation	
Neutralization reaction	

8. Briefly explain how the following tests work:

Test	Explanation
Radioimmunoassay (RIA)	
Enzyme-linked Immunosorbent Assay (ELISA)	
Western Blot	

9. List examples of each of the following kinds of serological reactions.

Type of reaction	Description
Precipitation reaction	
Agglutination reaction	
Immunofluorescence	
Complement fixation	
Neutralization reaction	

10. Indicate specific uses for the following test procedures.

Test	Use
Radioimmunoassay (RIA)	
Enzyme-linked Immunosorbent Assay (ELISA)	
Western Blot	

11. List four ways that cellular responses are used in immunological testing.

12. Matching the serological test with its description.

Answer	Serological test	Description
	1. Immunodiffusion	A. Antibodies are mixed with particulate antigens such as bacteria, fungi or red blood cells.
	2. Inmmunoelectrophoresis	B. Serum is mixed with known viral suspensions before the virus is used to infect a cell culture.
	3. Direct agglutination	C. Antigens and antibodies are allowed to diffuse through agar to form distinct lines of precipitate.
	4. Indirect agglutination	D. Enzyme reactions are used to label an antigen-antibody reaction.
	5. Hemagglutination Inhibition	E. Antigens are added to blood cells or latex beads before reacting with an antibody.
	6. Neutralization	F. This technique combines ELISA with electrophoresis to separate antigens.
	7. Direct fluorescence	G. Test organisms are fixed to a slide before reacting with a fluorescein-labeled antibody.
	8. Indirect fluorescence	H. Antibodies to viruses are detected by the inhibition of clumping of red blood cells.
	9. Radioimmunoassay (RIA)	I. Radioactively labeled material reacts with antibody bound to antigen.
	10. ELISA	J. This test combines precipitation with electrophoresis.
	11. Western blot	K. Fluorescent labeled anti-human gamma-globulin reacts with antigen-antibody complexes.

Self Test

1. The change from a negative serum without specific antibodies to serum positive for specific antibodies is called

   a. titer.
   b. seroconversion.
   c. rise in titer.
   d. serial dilution.
   e. None of the above.

2. In which of the following serological tests are visible antigen-antibody complexes formed?

   a. precipitation reactions
   b. agglutination reactions
   c. immunofluorescence tests
   d. complement fixation tests
   e. None of the above.

3. Which of the following tests combine electrophorphoresis with ELISA?

   a. radioimmunoassay (RIA)
   b. western blot
   c. complement fixation
   d. Both a and b
   e. None of the above.

4. Infectious organisms that are living but incapable of causing disease are said to be

   a. virulent.
   b. pathogenic.
   c. attenuated.
   d. inactivated.
   e. Such organisms do not exist.

5. A toxoid is a(n)

   a. type of antibody that combines with a toxin.
   b. type of enzyme that destroys toxins.
   c. inactivated toxin.
   d. antibody made against a toxin.
   e. artificial toxin.

6. Which of the following are tests used in cellular immunology?

   a. lymphocytes subset identification
   b. lymphocyte response to mitogens
   c. cytotoxic T cell function
   d. Both a and b
   e. a, b, and c

Match the statement with the type of immunity. Answers may be used once, more than once, or not at all. More than one answer may be used for a specific statement.

		Type of immunity
7.	The type of immunity produced after recovery from a disease	a. innate
8.	The type of immunity that protects humans from dog and cat distemper	b. naturally acquired active
9.	The type of immunity produced by the administration of gamma globulin for hepatitis	c. naturally acquired passive
10.	The type of immunity that is transferred from mother to fetus across the placenta	d. artificially acquired active
11.	The type of immunity results from the administration of the poliovirus vaccine	e. artificially acquired passive
12.	The type of immunity that is independent of previous exposure to foreign substances	
13.	The type of immunity that is not very long-lasting	
14.	The type of immunity does not involve the specific immune response	

Thought Questions

1. Explain how titer levels estimate an individual's protection against a specific disease.

2. Explain why recombinant vaccines are considered to be safer and better than other vaccines.

Answers to Self Test Questions

1-b, 2-a, 3-b, 4-c, 5-c, 6-e, 7-b, 8-a, 9-e, 10-c, 11-d, 12-a, 13-e, 14-a

Chapter 18   Immunologic Disorders

Overview

The immune system is very complicated with many components. When these components do not function properly, the immune system fails to respond in the appropriate manner. This produces disease. Hypersensitivities involve overreaction of the immune system. Autoimmune diseases result when the body reacts against substances of self. Immunodeficiency results in failure of the system to respond effectively against foreign antigens.

Learning Objectives

After studying the material in this chapter, you should be able to:

1. Define hypersensitivity and describe how it is produced.
2. List and describe the four kinds of hypersensitivities.
3. List the three types of localized anaphylactic reactions.
4. Describe the desensitization process.
5. Define autoimmunity and give examples of autoimmune diseases.
6. Explain how autoimmunity diseases may be treated.
7. Define immunodeficiency and give examples of immunodeficiency diseases.
8. Explain the difference between primary and secondary immunodeficiency.

Key Concepts

1. Hypersensitivity is an exaggerated reaction of the immune system to certain antigens called allergens.
2. Hypersensitivities can be classified in four groups based on the components of the immune system involved.
3. Autoimmunity occurs when the immune system responds to its own body antigens.
4. Immunodeficiency is the failure of the immune system to adequately respond to antigenic stimulation.
5. Immunodeficiencies may be primary, which are genetic or developmental in nature, or secondary, which are acquired.

Summary Outline

18.1 **Type I Hypersensitivities: Immediate IgE-mediated hypersensitivities**
   A.   **IgE attached to mast cells or basophils** reacts with specific antigen, resulting in the **release of powerful mediators** of the allergic reaction.
   B.   **Localized anaphylactic** (type I) reactions include
      1.   **Urticaria (hives)**
      2.   **Allergic rhinitis (hay fever)**
      3.   **Asthma**
   C.   **Generalized or systemic anaphylaxis** are rare, but serious and can lead to shock and death

    D.   **Immunotherapy or desensitization** is often effective in decreasing the type I hypersensitivity state. Engineered anti-IgE, is effective in treating asthma.

18.2 Type II Hypersensitivities: Cytotoxic hypersensitivities

    A.   Caused by antibodies that can destroy normal cells by **complement lysis** or by **antibody-dependent cellular cytotoxicity** (ADCC).

    B.   **Transfusion reactions**: The ABO blood groups have been the major cause of transfusion reactions.

    C.   **Hemolytic disease of the newborn**: The Rhesus blood groups are usually responsible for this disease.

18.3 Type III Hypersensitivities: Immune complex-mediated hypersensitivities

    A.   Mediated by small **antigen-antibody complexes** that activate complement and other inflammatory systems, attract neutrophils and contribute to inflammation.

    B.   Immune complexes cause **inflammatory disease** including **glomerulonephritis** and **arthritis**.

18.4 Type I Hypersensitivities: Delayed hypersensitivities

    A.   Depend on the actions of sensitized T lymphocytes.

    B.   **Tuberculin skin test**

    C.   **Contact hypersensitivities** occur in response to substances such as poison ivy, nickel in jewelry and chromium salts in leather products.

    D.   **Delayed hypersensitivity** is important in responses to many chronic, long-lasting infectious diseases.

18.5 Transplantation rejection of allografts is caused largely by Type IV cellular reactions.

18.6 Autoimmune diseases result from responses against self antigens and may be organ-specific or widespread.

    A.   Some autoimmune diseases are caused by **antibodies produced to body components** and others result from **cell-mediated reactions**.

    B.   Autoimmune diseases are usually **treated with drugs that suppress the immune and/or inflammatory responses**.

18.7 Immunodeficiency disorders may be primary **genetic** or **developmental** defects in any components of the immune response, or they may be **secondary** and **acquired**.

    A.   **Primary immunodeficiencies**

        1.   **B cell immunodeficiencies** result in diseases involving a lack of antibody production, such as agammaglobulinemias and selective IgA deficiency.

        2.   **T cell deficiencies** result.

    B.   **Secondary immunodeficiencies (acquired)** can result from malnutrition, immunosuppressive agents, infections (such as AIDS) and malignancies.

Terms You Should Know

Allergens	Delayed hypersensitivity	Primary immunodeficiency
Allergic rhinitis (hay fever)	Desensitization	Privileged sites
Allergies	Disseminated intravascular	Prophylaxis
Allografts	coagulation (DIC)	Rheumatoid arthritis
Anaphylaxis	Feeding tolerance	Secondary immunodeficiency
Antibody-dependent cellular	Hypersensitivities	Sensitized
cytotoxicity (ADCC)	Immune complex	Serum sickness
Arthus reaction	Immunotherapy	Shock
Asthma	Insulin-dependent diabetes	Systemic anaphylaxis
Autoimmune disease	mellitus	Urticaria (hives)
Contact dermatitis	Myasthenia gravis	

Learning Activities

1.   Define or identify the following:

Term	Definition
Hypersensitivity	
Autoimmunity	
Immunodeficiency	
Allergen	
Desensitization	
Anaphylaxis	
ADCC	
Hemolytic disease	
Transfusion	
Immune complex	

2.   List the name and major characteristics of the four kinds of hypersensitivities:

Hypersensitivity	Name	Characteristics
Type I		
Type II		
Type III		
Type IV		

3.   Describe the following autoimmune diseases:

Disease	Description
Lupus erythematosus	
Myasthenia gravis	
Insulin-dependent diabetes mellitus	
Rheumatoid arthritis	

4. Match the specific disorder with the type of hypersensitivity.

Answer		Type of hypersensitivity
	Transfusion reaction	Type I
	Hives	Type II
	Serum sickness	Type III
	Generalized anaphylaxis	Type IV
	Hemolytic disease of the newborn	
	Arthus reaction	
	Allergic rhinitis	
	Contact dermatitis	
	Disseminated intravascular coagulation	
	Tuberculin skin test	

5. Describe the mechanisms of causation for the following immune diseases.

Disease	Mechanism
Lupus erythematosus	
Myasthenia gravis	
Insulin-dependent diabetes mellitus	
Rheumatoid arthritis	

6. Define or describe the following types of immunodeficiency disease and list an example of each:

Type	Definition or description	Example
Primary		
Secondary		

7. Indicate the part of the immune system involved in these disorders.

Disease	Part of the immune system involved
DiGeorge Syndrome	
Chediak-Higashi disease	
Monoclonal gammopathy	
Congenital agammaglobulinemia	
Chronic granulomatous disease	
Acquired immunodeficiency syndrome (AIDS)	
Severe combined immunodeficiency	

Self Test

1. An exaggerated or inappropriate immune response is known as a(n)

   a. immunodeficiency.
   b. precipitation.
   c. histamine.
   d. hypersensitivity.
   e. allergen.

2. Localized anaphylaxis involving the skin is called

   a. allergic rhinitis.
   b. asthma.
   c. shock.
   d. hives
   e. hay fever.

3. Immunotherapy to prevent generalized anaphylaxis is done by injecting dilute doses of

   a. IgG antibodies.
   b. antihistamine.
   c. IgE antibodies.
   d. offending antigen.
   e. offending antibody.

4. Which of the following is most commonly involved in graft rejections?

   a. ABO antigens
   b. ABO antibodies
   c. MHC antigens
   d. MHC antibodies
   e. Both a and c are correct.

5. Which of the following would be a Type IV, or delayed hypersensitivity?

   a. hay fever
   b. allergic contact dermatitis
   c. glomerulonephritis
   d. penicillin reaction
   e. blood transfusion reaction

6. The symptoms of an immune complex reaction are due to

   a. an inflammatory response.
   b. histamines.
   c. autoimmune antigens.
   d. IgE.
   e. autoimmune antibodies.

7. Allergic contact dermatitis is mediated by

    a. sensitized T cells.
    b. IgG antibodies.
    c. IgE antibodies.
    d. basophils and mast cells.
    e. sensitized macrophages.

8. A hypersensitive reaction occurs

    a. during the first exposure to an antigen.
    b. in individuals with diseases of the immune system.
    c. on a second or subsequent exposure to an antigen.
    d. during autoimmune diseases.
    e. in immunologically deficient individuals.

9. Type I hypersensitivities

    a. are IgE mediated.
    b. have many symptoms that are due to histamine release.
    c. are caused by antibodies bound to basophils and mast cells.
    d. have symptoms that occur soon after exposure to the allergen.
    e. All of the above are true.

10. Hemolytic disease of the newborn occurs when an

    a. Rh-positive mother carries an Rh-negative fetus.
    b. O mother carries an AB fetus.
    c. Rh-negative mother carries an Rh-positive fetus.
    d. AB mother carries and O fetus.
    e. Two of the above are true.

11. Reaction of antigen with IgE antibodies attached to mast cells causes

    a. precipitation.
    b. complement fixation.
    c. degranulation.
    d. agglutination.
    e. asthma.

12. Immediate hypersensitivities are mediated by

    a. allergens.
    b. macrophages.
    c. humoral antibodies.
    d. antigens.
    e. T cells.

13. Immunotherapy for hypersensitivities probably works due to

    a. the production of IgG-blocking antibodies.
    b. development of suppressor T cells.
    c. saturation of IgE antibodies.
    d. All of the above are correct.
    e. Both a and b are correct.

14. Delayed hypersensitivities are mediated by

    a. allergens.
    b. macrophages.
    c. humoral antibodies.
    d. antigens.
    e. T cells.

15. Which of the following are true of immunodeficiency disorders?

    a. Primary immunodeficiencies are caused by genetic or developmental abnormalities.
    b. Primary immunodeficiencies are rare.
    c. Secondary immunodeficiencies are acquired.
    d. Secondary immunodeficiencies result form environmental factors.
    e. All of the above are true.

16. Which of the following are primary immunodeficiency diseases?

    a. Severe combined immunodeficiency (SCID)
    b. CD3 deficiency
    c. AIDS
    d. Only a and b are correct.
    e. a, b, and c are correct.

17. Which of the following are secondary immunodeficiency diseases?

    a. DiGeorge syndrome
    b. Monoclonal gammopathy
    c. Chediak-Higashi disease
    d. Chronic granulomatous disease
    e. Both b and c.

18. Autoimmune diseases occurs because the immune system

    a. recognizes host cells as foreign.
    b. recognizes host cells as self.
    c. fails to recognize host cells as foreign.
    d. fails to recognize host cells as self.
    e. dislikes your car.

19. Which of the following are autoimmune diseases?

    a.   myasthenia gravis
    b.   insulin-dependent diabetes mellitus
    c.   rheumatoid arthritis
    d.   Both a and c are correct.
    e.   a, b, and c are correct.

20. Which of the following are true of feeding tolerance?

    a.   It is a phenomenon in which an antigen introduced by the oral route can cause a local intestinal immune response.
    b.   It is a phenomenon in which an antigen introduced by the oral route can suppress a response when the same antigen is later introduced by other routes.
    c.   It is currently being used to treat rheumatoid arthritis.
    d.   Both a and b are correct.
    e.   a, b, and c are correct.

Thought Questions

1.   Describe how degranulation occurs.

2.   Describe the diagnosis and treatment of type I hypersensitivities..

Answers to Self Test Questions

1-d, 2-d, 3-d, 4-c, 5-b, 6-a, 7-a, 8-c, 9-e, 10-c, 11-c, 12-c, 13-e, 14-e, 15-e, 16-d, 17-b, 18-a, 19-e, 20-d

Chapter 19   Host-Microbe Interactions

Overview

When microorganisms and human hosts meet, the effect of microorganisms on the individual will depend on how the microorganism and host interact. Sources of microorganisms may include normal microbial flora, transient microbial flora or pathogens introduced into the body from soil, water, animals and other people. Infection occurs when the microorganisms grow and multiply on or in the body of the host. Infections may produce disease if the microbes are virulent. Toxins produced by microorganisms may also cause disease. Important steps in the pathogenesis of infectious diseases are adherence to body cells or tissues, colonization and invasion either between or through cells, and interference with phagocytosis and complement activities. If the host defense mechanisms are not adequate to protect the host or the microbes are able to evade the defense mechanisms, disease occurs. Thus, disease is dependent not only on the presence of a microorganism but also on how the host interacts with that microorganism. This chapter examines the relationship between microorganism and host.

Learning Objectives

After studying the material in this chapter, you should be able to:

1. Define the following:
   - Pathogen
   - Pathogenicity
   - Opportunistic pathogen
   - Infection
   - Disease
   - Primary infection
   - Secondary infection
2. Define normal flora and list five reasons why it is important.
3. Define symbiosis and differentiate among three types of symbiosis.
4. Define
   - Virulence
   - Communicable disease
   - Symptoms
   - Signs
   - Carrier
5. Describe the course of an infectious disease including the following periods:
   - Incubation period
   - Period of illness
   - Convalescence
6. Define or identify:
   - Acute disease
   - Chronic disease
   - Localized disease
   - Systemic disease

- Inapparent disease
- Latent disease
- Viremia
- Bacteremia
- Septicemia
- Toxemia

7. List Koch's postulates and describe how they are applied.
8. Explain how the molecular postulates are used.
9. Explain how microbial diseases are transmitted.
10. Define the following terms:
    - Fomites
    - Portal of entry
    - Portal of exit
11. List the major portals of entry
12. List the major portals of exit.
13. Define the following terms:
    - Adherence
    - Colonization
    - Quorum sensing
14. List the major virulence factors of microorganisms and explain how they cause disease.
15. Define:
    - Toxin
    - Endotoxin
    - Exotoxin
    - Superantigen
16. Describe the role of cytolytic toxins and toxic enzymes, using specific examples.
17. List specific examples of organisms that produce exotoxins.
18. List specific examples of organisms that contain endotoxins.
19. Describe how superantigens function.
20. Describe the role that bacteriophages play in the pathogenicity of some bacteria.
21. Define pathogenicity island and explain its role in the genetic transfer of virulence factors.
22. Explain how fungi cause damage to host cells.
23. Explain how protozoans cause damage to host cells.
24. Describe how viruses cause disease and evade host defense mechanisms.
25. Explain how the host defense mechanisms, host genetic constitution and host physiology can contribute to the etiology (causation) of disease.
26. List and describe the means by which bacteria evade the host defenses.

Key Concepts

1. Normal flora consist of the microorganisms that grow in or on the body without producing harmful effects under normal circumstances.
2. Infection occurs when parasitic microorganisms grow in or on the body of the host.
3. Normal flora is dynamic and changes in response to the host environment.
4. Pathogens are microorganisms that produce disease.
5. Virulence describes the factors that are responsible for the ability of a microorganism to produce disease.
6. Infectious diseases are communicable and are spread between people or animals.

7. The stages of infectious diseases are the incubation period, the period of illness, and the convalescent period.
8. Infections can be acute or chronic, localized or systemic, inapparent, or latent.
9. Koch's Postulates are used with organisms that can be cultured in the laboratory to demonstrate that cause of an infectious disease. The Molecular Postulates use genes to identify virulence factors in pathogenic strains of microorganisms.
10. Infectious agents are transmitted from soil, water, animals, and normal flora by direct or indirect contact.
11. Pathogenesis of infectious diseases is due to adherence of the microbe to host cells or tissues, colonization and invasion between or through cells, and interference with phagocytosis and complement activities. Toxins may be the cause of damage to host cells.
12. Genes that code for virulence factors may be transferred between bacteria by transduction, transformation, and conjugation.
13. Algae, fungi, protozoa, multicellular parasites, and viruses each have multiple ways of causing disease.
14. Host responses may contribute to disease. The outcome of an infection depends on both the activities of the microbe and the host response.
15. Microorganisms may evade host responses in many various ways including genetic variation of their virulence factors, escaping immune responses by mimicking host substances, degrading host antibodies, taking over the host cell actin, and escaping destruction by phagocytes.

Summary Outline

19.1 The skin and mucous membranes constitute an anatomical barrier to invading microorganisms and supply the foundation a complex **ecosystem**.
19.2 **Normal flora** is comprised of the microorganisms that grow in or on the body without producing obvious harmful effects.
    A. **Symbiosis** describes the living together of two dissimilar organisms.
        1. **Commensalism**: One partner benefits, the other is unaffected.
        2. **Mutualism**: Both partners benefit in mutualism.
        3. **Parasitism**: The parasite benefits at the expense of the host.
    B. The **normal flora** is **acquired at birth** and **changes** in response to **variations in the environment**, such as changes in diet, acidity, or antibiotic intake.
    C. **Normal flora inhibits potentially harmful organisms** by **preventing attachment**, **competing for essential nutrients, producing antimicrobial substances, stimulating the immune system** and **inducing** the production of **antibodies** that cross-react with potential pathogens.
19.3 Principles of infectious disease
    A. Colonization of a host
        1. **Infection** occurs when parasitic organisms grow in or on the host.
        2. **Disease** occurs if the infection causes damage.
        3. **Symptoms** are the effects of the disease reported by the patient.
        4. **Signs** are effects of the disease that can be observed.
        5. **A primary infection** is the initial infection.
        6. **A secondary infection** occurs as a result of the primary infection.
    B. Pathogenicity
        1. **Pathogens** are **disease-producing organisms.**
        2. **Opportunist** cause disease when the body's defense systems are compromised
        3. **Virulence**, the degree of pathogenicity, describes the properties (**virulence factors**) of an infectious agent that promote its pathogenicity.

C. **Infectious diseases** are **communicable** or **contagious** and spread between people or animals.
D. **Stages** of infectious diseases
   1. **Incubation period**
   2. **Period of illness**
   3. **Convalescence**
E. **Infections** may be **acute** or **chronic,** or **latent.** Infections may be **localized** or **systemic**.

19.4 **Establishing the cause** of an infectious disease
   A. **Koch's postulates** are used with organisms that can be cultured in the laboratory to demonstrate the cause of an infectious disease.
   B. **Molecular postulates** use genes to identify virulence factors in pathogenic strains of organisms.

19.5 Establishment of infection
   A. **Adherence**: Microorganisms adhere to host cell receptors by means of adhesins.
   B. **Colonization**: Microorganisms colonize the host in order to become established.
   C. Type III secretion systems of Gram-negative bacteria allow them to deliver compounds directly to the host cells.

19.6 **Invasion of tissues**: A pathogen may enter through a break in the skin or mucous membranes, destroy the basement membrane underlying epithelial cells, or go between or through the cells of the membrane. Some may be endocytosed or phagocytized.

19.7 Avoiding the host defenses
   A. **Pathogens** make a variety of substances that **inhibit phagocytosis** and **complement** activities.
   B. Some bacteria can evade host defenses by remaining inside of host cells.
   C. Some **serum-resistant** bacteria can interfere with the activation of the complement system via the alternative pathway, thus postponing the formation of the membrane attack complex.
   D. Bacterial mechanisms can prevent encounters with phagocytes and destroy phagocytic cells.
   E. Bacterial means of avoiding recognition and attachment by phagocytes include **capsules, M protein, Mac proteins,** and **Fc receptors.**
   F. Mechanisms for the bacteria to survive within the phagocyte include:
      1. Escape from the phagocyte
      2. Preventing phagosome-lysosome fusion
      3. Surviving within the phagosome
   D. Mechanisms to avoid antibodies include **IgA protease, antigenic variation**, and mimicking the host.

19.8 Damage to the host
   A. **Exotoxins** are powerful toxic proteins with specific damaging effects. They may be classified as **neurotoxins, enterotoxins,** or **cytotoxins** based on the kinds of host cells that they damage.
   B. **Superantigens** bind directly to many T cells at sites distinct from the regular antigen receptor sites. This causes the **release of** huge amounts of **cytokines** that have **toxic effects** in such large quantities.
   C. **Endotoxins of Gram-negative cell walls** consist of lipopolysaccharides that contain a **toxic lipid A portion** and an **O polysaccharide antigen**. **Lipid A** is responsible for the toxic properties that include **fever** and sometimes **shock**.
   D. **Peptidoglycan** and some other bacterial components induce various cells to produce proinflammatory **cytokines**.
   F. **Cytolytic toxins** and **toxic enzymes** break down cells or tissue components.
   G. **Antigen-antibody complexes** can cause damage to the kidneys and joints; **cross-reactive antibodies** can promote an autoimmune response.

19.9   Mechanisms of viral pathogenesis
    A.   Binding to host cells and invasion
        1.   Some viruses bind to and infect mucous membrane cells; others enter at sites that are damaged or penetrated.
        2.   Viruses attach to specific receptors on the target cells.
    B.   Avoiding immune responses
        1.   Some viruses can avoid the effects of interferon.
        2.   Some viruses can regulate **apoptosis** of the host cell.
        3.   Some viruses transfer directly from cell to cell in order to avoid antibodies.
        4.   The surface antigens of some viruses change more quickly than the production of complimentary antibodies.
    C.   Viruses and host damage: Viruses can damage the host by:
        1.   Direct damage
        2.   Virally induced apoptosis
        3.   Altering the immune response to the infection
19.10   Mechanisms of eukaryotic pathogenesis
    A.   Fungi: Saprophytic fungi are usually opportunistic; dermatophytes cause superficial infections of the skin, hair, and nails. The dimorphic fungi cause the most serious fungal infections.
    B.   Eukaryotic parasites attach to specific receptors on host cells. They use a variety of mechanisms to avoid antibodies; the damage varies considerably.

Terms You Should Know

A-B toxin	Koch's postulates	Portal of exit
Antitoxin	Latent infection	Primary infection
Bacteremia	Leukocidins	Quorum sensing
Chronic infection	Lipid A	Secondary infection
Colonization	Localized infection	Septicemia
Commensalism	Molecular postulates	Signs
Communicable disease	Mutualism	Subclinical (inapparent) infection
Convalescence period	Normal flora	Superantigen
Disease	O antigen	Symbiosis
Endotoxin	Opportunistic pathogen	Symptoms
Exotoxin	Parasite	Systemic infection
Fomite	Parasitism	Toxemia
Host	Pathogen	Viremia
Incubation period	Pathogenicity	Virulence
Infection	Pathogenicity island	
Infectious dose	Portal of entry	

Learning Activities

1. Define the following terms:

Term	Definition
Normal flora	
Symbiosis	
Commensalism	
Mutualism	
Parasitism	
Host	

2. List five roles of normal flora:

1.	
2.	
3.	
4.	
5.	

3. List and differentiate among the three types of symbiotic relationships in the following table:

Type	Benefit to microbe	Benefit or harm to host?

4. Define the following terms:

Term	Definition
Disease	
Pathogenicity	
Pathogen	
Opportunistic pathogen	
Virulence	
Avirulent	
Communicable disease	

5. Differentiate between the following kinds of infections:

Primary infection	
Secondary infection	

6. Define the following:

Colonization	
Infection	
Adherence	

7.   Explain what happens during the following periods of an infectious disease.

Period	Events
Incubation	
Illness	
Convalescence	
Latent	

8.   Define or identify the following:

Term	Definition
Symptoms	
Signs	
Acute infection	
Chronic infection	
Localized infection	
Systemic infection	
Viremia	
Bacteremia	
Septicemia	
Toxemia	
Subclinical infection	
Latent infection	

9. List the four Koch's postulates that identify a microorganism as the cause of a specific disease.

1.	
2.	
3.	
4.	

10. List the four molecular postulates.

1.	
2.	
3.	
4.	

11. Describe how the molecular postulates are used.

12. List four major sources of infectious agents.

1.	
2.	
3.	
4.	

13. Define the following terms:

Terms	Definition
Fomites	
Portal of entry	
Portal of exit	

14. List the six steps essential for the virulence and survival of pathogens.

1.	
2.	
3.	
4.	
5.	
6.	

15. Define the following:

Term	Definition
Endotoxin	
Exotoxin	
Antitoxin	
Toxoid	

16. Differentiate between exotoxins and endotoxins by completing the following table.

Characteristic	Exotoxins	Endotoxins
Bacterial source		
Heat-labile		
Toxicity		
Toxoids		
Formation of antibodies		

17. Describe the role of each of the following molecules in the production of disease.

Molecule	Role in the production of disease
Protein G (*S. pyogenes*)	
Protein A (*S. aureus*)	
Leukocidins	
Superantigens	
Phospholipase	
Hemolysin	
Streptokinase	
Collagenase	
Hyaluronidase	
Lipase	
Deoxyribonuclease	
Alpha toxin	

18. List four ways in which bacteria try to evade the host defenses.

1.	
2.	
3.	
4.	

19. The following are virulence factors of microorganisms. Describe how each helps a microbe escape destruction and/or cause disease.

	Virulence factor	How they work
A	Capsules	
B	Protein A	
C	Waxes	
D	Leukocidins	
E	Hemolysins	
F	Coagulases	
G	Kinases (Streptokinase and Staphylokinase	
H	Invasive enzymes (Hyaluronidase and collagenase)	

20. The following is a list organisms that produce toxins. Indicate which produce endotoxins and which produce exotoxins.

Microbe	Type of toxin
Streptococcus pyogenes	
Clostridium botulinum	
Salmonella typhi	
Corynebacterium diphtheriae	
Neisseria meningitidis	
Vibrio cholerae	
Staphylococcus aureus	
Proteus vulgaris	

21. List the three general types of exotoxins and give the general mode of action (how does it affect the cells) of each type:

	General type	**Mode of action**
1.		
2.		
3.		

Self Test

1. The condition in which the homeostasis of the body is impaired is called

   a. infection.
   b. colonization.
   c. disease.
   d. All of the above.
   e. None of the above.

2. The presence of microorganisms in a place where they are not usually found is called a(n)

   a. infection.
   b. colonization.
   c. disease.
   d. All of the above.
   e. None of the above.

3. The symbiotic relationship called parasitism is one in which

   a. the host is benefited and the microbe is benefited.
   b. the host is harmed and the microbe is benefited.
   c. the host is neither harmed nor benefited, but the microbe is benefited.
   d. the host is benefited and the microbe is harmed.
   e. neither the host nor the microbe is benefited or harmed.

4. Which of the following do NOT contribute to a pathogen's virulence?

   a. endotoxins
   b. capsules
   c. collagenase
   d. exotoxins
   e. All of the above contribute to virulence.

5. Endotoxins

   a. all have the same effect.
   b. are found in both Gram-negative and Gram-positive bacteria.
   c. are part of the Gram-positive cell wall.
   d. are heat-labile.
   e. are proteins.

6. Opportunistic pathogens are

   a. the principal pathogens to infect a host.
   b. always the cause of secondary infections.
   c. microbes that cause disease when introduced into an unusual location or into an immunologically compromised host.
   d. infections that are widespread through the body.
   e. infections that are localized within the body.

7. A primary infection differs from a secondary infection in that

   a. a primary infection is more severe than a secondary infection.
   b. a primary infection is less severe than a secondary infection.
   c. a primary infection is always due to an opportunistic pathogen.
   d. a primary infection is one that occurs in a previously healthy person.
   e. a primary infection is one that occurs as the result of another infection.

8. Which of the following is false for Koch's postulates?

    a. The microorganism must be present in every case of the disease.
    b. The microorganism must be grown in pure culture from the diseased host.
    c. The same disease must be produced when a pure culture of the microorganism is introduced into a susceptible host.
    d. The microorganism must be recovered from the experimentally infected host.
    e. All of the above are true.

9. Which of the following are factors in the pathogenesis of infectious diseases?

    a. Transmission of the causative agent to a susceptible host.
    b. Adherence of the agent to a target tissue.
    c. Colonization and invasion of host tissues or cells.
    d. Damage to the host by the activities of toxins or other mechanisms.
    e. All of the above.

10. Which of the following are host responses that contribute to pathogenesis?

    a. hypersensitivity
    b. impaired body defenses
    c. genetic factors
    d. age and stress
    e. All of the above are correct.

Thought Questions

1. Discuss the importance of normal flora in preventing disease.

2. Explain the significance of Koch's work.

Answers to Self Test Questions

1-c, 2-a, 3-b, 4-e, 5-a, 6-c, 7-d, 8-e, 9-e, 10-e

Chapter 20   Epidemiology

Overview

Epidemiology, the basic science of public health, is the study of the cause and distribution of health states, including diseases, in populations. Through the identification of causative (etiological) agents, sources and modes and routes of transmission of disease, intervention and prevention of disease may be accomplished. Types of epidemiological studies are presented as well as a summary of current disease surveillance networks. Emerging diseases and nosocomial infections are also covered.

Learning Objectives

After studying the material in this chapter, you should be able to:

1. Define epidemiology
2. Define the following:
    -   Portal of exit
    -   Portal of entry
    -   Communicable diseases
    -   Non-communicable diseases
3. List the major portals of exit and portals of entry.
4. Define and differentiate between morbidity rate and mortality rate.
5. Define and differentiate between incidence and prevalence.
6. Define
    -   Endemic
    -   Epidemic
    -   Outbreak
    -   Pandemic
7. Define reservoir and give examples.
8. Define zoonoses.
9. Differentiate between horizontal and vertical transmission of disease.
10. List specific mechanisms for the transmission of microbial diseases.
11. Define
    -   Fomites
    -   Droplet nuclei
    -   Vectors
12. Differentiate between mechanical and biological vectors.
13. List and describe the major factors that influence the epidemiology of disease.
14. Differentiate among descriptive, analytical and experimental epidemiological studies.
15. Define
    -   Risk factor
    -   Placebo
    -   Double-blind study

16. Describe the following epidemiological approaches
    - Cross-sectional
    - Retrospective or Case-Control
    - Prospective
17. Differentiate between observational and experimental approaches to scientific investigation.
18. Describe how infectious disease surveillance is carried out.
19. Define or explain the following terms:
    - Centers for Disease Control and Prevention (CDC)
    - Morbidity and Mortality Weekly Report (MMWR)
    - Notifiable diseases
    - World Health Organization (WHO)
    - Emerging diseases
20. List and describe some of the factors that contribute to the emergence and reemergence of diseases.
21. Define a nosocomial infection and explain why they occur.
22. Describe how nosocomial infections can be prevented.

Key Concepts

1. Epidemiology is the science that deals with cause and distribution of disease in human populations.
2. Epidemiology seeks to prevent disease through the determination of risk factors, reservoirs and modes of transmission, and the intervention on these factors.
3. Epidemiology is concerned with the rates of diseases in populations rather than the course of disease in an individual.
4. Preventing susceptible people from coming in contact with the reservoir can prevent infectious disease.
5. Reservoirs of disease can include people, animals or the environment.
6. Infectious microorganism must exit an infected host in order to spread a disease. The intestinal tract, the respiratory tract, the genital tract, and the urinary tract are common portals of exit.
7. Diseases can be transmitted by direct or indirect contact. Specific paths of transmission involve food, air, water, and arthropod vectors.
8. The portal of entry of a pathogen can effect outcome of disease including the type and severity of the disease.
9. Population characteristics such as prior exposure or vaccination to a disease, age, gender, religion, cultural practices, general health, and genetic constitution can influence susceptibility to disease.
10. There are two kinds of epidemiological studies (a) descriptive that attempt to identify risk factors that lead to disease and (b) analytical that attempt to determine which factors are associated with the disease. Experimental studies are used to determine the effectiveness of a treatment or a preventative technique.
11. It may be necessary to distinguish between different strains within a species in order to identify the source of an epidemic.
12. Infectious disease surveillance is done by the Center for Disease Control and Prevention, as well as state and local health departments, in order to detect the trends of disease.
13. While humans have been successful in eliminating some infectious diseases, microorganisms have been very successful in taking advantage of new opportunities in which to thrive and multiply.
14. Nosocomial infections are hospital acquired infections that may originate from other patients, the hospital environment, medical personnel, or the patient's own normal flora.

Summary Outline

20.1   Principles of epidemiology

    A. **Epidemiology** is the study the **frequency** and **distribution** of disease in order to identify its **cause, source** and **route of transmission**.

       1. **Epidemiology** is concerned with the **rate of disease in a given population.**

       2. **Endemic diseases** are constantly present in a population.

       3. An **epidemic** is the occurrence of a disease in larger numbers than would be expected in a given population.

       4. An **outbreak** is a cluster of cases occurring during a specific brief time period and affecting a specific population.

       5. A **pandemic** is an epidemic of world-wide proportions.

    B. **Reservoirs of infection**: Continuous source

       1. **Human reservoirs: Infected people** and people who are **carriers** of the infectious agent

       2. **Nonhuman animal reservoirs: Zoonotic diseases** are those such as plague and rabies that can be transmitted to humans but exist primarily in other animals.

       3. Diseases with **environmental reservoirs** are probably impossible to eliminate.

    C. **Portals of exit** include

       1. In **feces**

       2. In **respiratory droplets**

       3. On **skin** cells

       4. In **genital secretions**

       5. In **urine**

    D. Transmission

       1. **Hand washing** is an important control measure in preventing diseases that are spread through **direct** or **indirect contact**, as well as those that spread via **contaminated food**.

       2. **Direct contact** occurs when one person physically touches another. Diseases with a **low infectious dose** and those caused by **pathogens that cannot survive for extended periods** in the environment are generally transmitted through **direct contact**.

       3. **Indirect contact** involves transfer of pathogens via **fomites**.

       4. **Droplet transmission** of respiratory pathogens is considered direct contact because of the close proximity involved.

       5. **Food-borne pathogens** can originate from the animal reservoir or from contamination during food preparation.

       6. **Waterborne pathogens** often originate from sewage contamination.

       7. **Droplet nuclei**, dead skin cells, household dust and soil may carry airborne respiratory pathogens. This type of transmission is most difficult to control.

    E. **Vectors**

       1. **Mechanical vectors** result in moving the microbe from one place to another.

       2. **Biological vectors** are a required part of a parasite's life cycle.

       3. Prevention of vector-borne disease relies on control of the vector.

    F. The **portal of entry** of a pathogen can effect the outcome of disease.

    G. **Factors** that influence the epidemiology of disease

       1. The **dose**: The probability of infection and disease is lower if an individual is exposed to fewer numbers of pathogens.

       2. Diseases with a long **incubation period** can spread extensively before the first cases appear.

3. Population characteristics
   a. **High percentages of immunity** in a population make it difficult for a disease to spread.
   b. **Malnutrition, overcrowding** and **fatigue increase the susceptibility** of people to infectious diseases.
   c. **Age**: The **very young** and the **elderly** are generally **more susceptible** to infectious agents.
   d. **Natural immunity** can vary with **genetic background**, but it is difficult to determine the relative importance of genetic, cultural and environmental factors.

20.2 **Epidemiological studies**
A. **Descriptive studies** attempt to identify **potential risk factors** that correlate with the development of disease by creating a profile of the persons who became ill. Three types of variables describe the population: **Person, place and time**.
B. **Descriptive studies** attempt to identify **potential risk factors** that correlate with the development of disease by creating a profile of the persons who became ill. Three types of variables describe the population: **Person, place and time**.
C. **Analytical studies** try to determine which risk factors are actually relevant to disease development.
   1. A **retrospective study** compares the activities of cases with controls to determine the cause of the epidemic.
   2. A **prospective study** looks ahead, comparing **cohort** groups, to determine if the identified risk factors predict a tendency to develop disease.
D. **Experimental studies** are generally used to evaluate
   1. The **effectiveness of a treatment**
   2. **Intervention** in preventing disease.

20.3 Infectious disease surveillance
A. The **Centers for Disease Control and Prevention** collects data on diseases of public health importance and summarizes their status in the **Morbidity and Mortality Weekly Report**.
B. **State public health departments** are involved in infection surveillance and control.
C. The **World Health Organization (WHO)** is an international agency devoted to achieving the highest possible level of health for all people.

20.4 Trends in disease
A. Reduction and eradication of disease
   1. Smallpox has been eradicated.
   2. The World Health Organization hopes to soon eliminate other diseases including dracunculiasis, polio and measles.
B. Emerging diseases include those that are new or newly recognized and familiar ones that are reemerging after years of decline.

20.5 **Nosocomial infections**: Hospital acquired infections
A. **Reservoirs** of infectious agents in hospitals include other **patients**, the **hospital environment, medical personnel** or the **patient's own flora**.
B. **Transmission of infectious agents in hospitals** can occur by **diagnostic and therapeutic procedures** and **health care personnel.**
C. **Hand-washing** is an important means of prevention.

Terms You Should Know

Analytical study
Biological vector
Carrier
Cohort group
Common-source epidemic
Communicable disease
Cross-sectional study
Descriptive study
Direct contact
Double-blind
Droplet nuclei
Droplet transmission
Emerging diseases
Endemic
Epidemic

Epidemiology
Experimental study
Fecal-oral transmission
Fomite
HEPA filter
Incidence
Index case
Indirect contact
Mechanical vector
Morbidity rate
Mortality rate
Non-communicable disease
Nosocomial infections
Notifiable diseases
Outbreak

Pandemic
Placebo
Portal of entry
Portal of exit
Prevalence
Propagated epidemic
Prospective study
Reservoir
Retrospective study
Risk factors
Standard precautions
Universal precautions
Vector
Zoonotic disease

Learning Activities

1. Differentiate between morbidity rate and mortality rate.

2. Differentiate between prevalence and incidence.

3. Differentiate among endemic, epidemic, outbreak and pandemic.

4. Differentiate between mechanical vector and biological vector.

5. List the four major portals of entry:

1.	
2.	
3.	
4.	

6. Indicate the order of frequency of the following major portals of exit:
   1 = most frequent, 4 = least frequent

	Frequency	Portal of exit
A		Urogenital tract
B		Wound infections
C		Respiratory tract
D		Gastrointestinal tract

7. List four examples of a reservoir of infection.

1.	
2.	
3.	
4.	

8. List six ways in which microbial diseases can be transmitted.

1.	
2.	
3.	
4.	
5.	
6.	

9. List the three types of variables that epidemiologists collect when they conduct a cross-sectional study.

1.	
2.	
3.	

10. List eight factors that can influence the epidemiology of a disease.

1.	
2.	
3.	
4.	
5.	
6.	
7.	
8.	

11. Define the following:

Term	Definition
Descriptive epidemiological studies	
Analytical epidemiological studies	
Experimental epidemiological studies	

12. Compare and contracts the following epidemiological techniques used in analytical epidemiology.

Cross-sectional	Prospective	Case-Control

Self Test

1. The study of the occurrence, transmission, causes, and distribution of diseases in populations is called

   a. public health.
   b. epidemiology.
   c. ecology.
   d. etiology.
   e. diagnostic microbiology.

2. A nosocomial infection is an infection

   a. that can come from the patient's own normal flora.
   b. acquired during hospitalization.
   c. that is always caused by medical personnel.
   d. Both a and b are correct.
   e. Both b and c are correct.

3. Which of the following can be reservoirs for human infections?

   a. food and water
   b. other humans
   c. animals
   d. All of the above are reservoirs for human infections.
   e. Only b and c are reservoirs.

4. A simple method for interrupting the person-to-person transmission of disease is

   a. immunization.
   b. hand-washing.
   c. prophylactic chemotherapy.
   d. isolation.
   e. quarantine.

5. The most common source of disease for people is

   a. pets.
   b. water.
   c. other people
   d. food.
   e. air.

6. Fomites are

   a. insect vectors.
   b. animate objects.
   c. an ancient tribe of Israel.
   d. inanimate objects.
   e. biological vectors.

7. Droplet nuclei are significant in the transmission of disease of the

   a. digestive system.
   b. reproductive system.
   c. nervous system.
   d. respiratory system.
   e. skin.

8. Descriptive epidemiological studies classify populations of people by which of the following characteristics?

   a. person
   b. place
   c. time
   d. occupation
   e. a, b and c are correct.

9. The type of epidemiological study done to determine the prevalence of characteristics that may be associated with a particular disease is

   a. cross-sectional.
   b. retrospective.
   c. prospective.
   d. experimental.

10. Which of the following are contributing to emerging diseases?

   a. microbial evolution
   b. complacency and the breakdown of public health infrastructure
   c. advances in technology
   d. climate changes
   e. All of the above are true.

Thought Questions

1. Describe what might be done to drastically reduce the spread of nosocomial infections.

2. There is an outbreak of a microbial disease that has not been previously seen. What kinds of studies would you do to help control this disease and why?

3. After many years the incidence of malaria is increasing. Explain why this disease is reemerging.

Answers to Self Test Questions

1-b, 2-d, 3-d, 4-b, 5-c, 6-d, 7-d, 8-e, 9-a, 10-c

Chapter 21   Antimicrobial Medications

Overview

Since ancient times we have sought a means for the treatment of infectious disease. While chemicals may have been previously used to treat infectious diseases, they were not documented or applied in a scientific manner until Paul Ehrlich used Salvarsan to treat syphilis. With this event and the discovery of penicillin by Alexander Fleming, the chemotherapy of infectious disease was born. Antimicrobial agents work by interfering with a microbe's metabolism. Many different antimicrobial agents are available, falling into categories based on the type of microbe on which they can be used and their mode of action. This chapter presents the antimicrobial medications, their activity and their applications. Resistance to antimicrobials is also discussed.

Learning Objectives

After studying the material in this chapter, you should be able to:

1. Identify the first successful antimicrobial agent and the scientist who discovered it.
2. Identify the first antibiotic discovered and the scientist who discovered it.
3. Define or identify
    *   Chemotherapeutic agent
    *   Antimicrobial drug or agent
    *   Antibiotic
    *   Selective toxicity
    *   Therapeutic index
    *   Bacteriostatic
    *   Bactericidal
4. Explain why one must consider the following in the selection of a specific antimicrobial agent to treat an individual with a specific disease.
    *   Selective toxicity
    *   Spectrum of activity
    *   Tissue distribution
    *   Metabolism and excretion of the drug
    *   Adverse effects
    *   Synergistic combinations
    *   Microbial resistance
5. Differentiate between broad-spectrum and narrow-spectrum antimicrobials.
6. Relative to the use of combinations of antimicrobial drugs explain the following terms:
    *   Synergistic
    *   Antagonistic
    *   Additive
7. List and describe the three major types of adverse effects caused by antimicrobial agents.
8. Explain why microbial resistance to antimicrobial agents is a major problem.
9. Explain why antimicrobial agents are more effective against bacteria than against the eukaryotic pathogens and viruses.

10. List the major antibacterial drugs by the following modes of action and include the important uses of these drugs.
    - Inhibition of cell wall synthesis
    - Inhibition of protein synthesis
    - Inhibition of nucleic acid synthesis
    - Inhibition of metabolic pathways
    - Interference of cell membrane function
    - Interference of *Mycobacterium tuberculosis* metabolism
11. Identify the β-lactam drugs.
12. Explain the action of β-lactamase or penicillinase.
13. Explain how the sensitivity of bacteria to a specific antibacterial agent can be determined in the laboratory.
14. Briefly describe the following tests:
    - Determination of minimum inhibitory concentration (MIC)
    - Determination of minimum bactericidal concentration (MBC)
    - Diffusion assay
    - Kirby-Bauer disc diffusion
    - E test
15. Be able to read a standard curve that correlates the size of the zone with the concentration of antibiotic in the diffusion assay test. (See Perspective 21.1.)
16. List four mechanisms of antimicrobial drug resistance.
17. Describe the causes of antimicrobial drug resistance.
18. Differentiate between vertical and horizontal evolution in the acquisition of resistance.
19. Explain what is being done to slow the spread of antimicrobial drug resistance.
20. List the major antiviral drugs by the following modes of action and include the important uses of these drugs.
    - Blocking viral uncoating
    - Inhibition of viral nucleic acid synthesis
    - Inhibition of viral replication and assembly
21. List the major antifungal drugs by the following modes of action and include the important uses of these drugs.
    - Disrupts or damages the cell membrane
    - Inhibition of nucleic acid synthesis
22. List the major antiprotozoan drugs and antihelminth drugs by the following general uses and give mode of action if known.
    - Treatment of intestinal protozoa
    - Treatment of *Plasmodium* and *Toxoplasma*
    - Trypanosomes and *Leishmania*
    - Intestinal and tissue helminths

Key Concepts

1. Antimicrobial agents are chemicals that are effective in inhibiting the growth and/or metabolism of microbes infecting an individual.
2. Antibiotics are natural antimicrobial chemicals that are produced by some microorganisms.
3. Chemotherapeutic agents are selectively toxic to some microorganisms while not affecting or minimally affecting other microbes or the host.

4. Selective toxicity is based on the differences in metabolism and structure between eukaryotic and prokaryotic cells

5. Viruses and eukaryotic pathogens such as fungi, protozoans and helminths have fewer targets than bacteria for selective toxicity.

6. Mutations and the transfer of information among microbes have made resistance to antimicrobial agents a problem.

7. Laboratory tests of the susceptibility of microbes to specific antimicrobial agents have aided the prediction of the effectiveness of the agent in a diseased host.

Summary Outline

21.1   History and development of antimicrobial drugs
  A.   Definitions
    1.   **Chemotherapeutic agents** are chemicals used as therapeutic drugs.
    2.   **Antimicrobial drugs** (antimicrobials) are chemotherapeutic agents that are effective against microbial infections.
    3.   **Antibiotics** are antimicrobial chemicals that are naturally produced by microorganisms.
  B.   The development of **Salvarsan** by Paul Ehrlich was the first documented example of an antimicrobial medication.
  C.   Alexander Fleming discovered that the fungus *Penicillium* produces **penicillin** that kills some bacteria.
  D.   **Antimicrobial drugs** can be **chemically modified** to give them new properties. **Penicillin** has been altered to create a **family of drugs** with a variety of new characteristics.
21.2   Features of antimicrobial drugs
  A.   Most modern **antibiotics** come from the bacteria *Streptomyces* and *Bacillus* and the eukaryotic fungi *Penicillium* and *Cephalosporium*.
  B.   Medically useful antimicrobials are **selectively toxic**. The relative toxicity of a drug is expressed as the **therapeutic index**, which is the lowest dose toxic to the patient divided by the dose typically used for therapy.
  C.   Antimicrobial action
    1.   **Bacteriostatic drugs inhibit the growth of microorganisms**.
    2.   **Drugs that kill microorganisms are bactericidal**.
  D.   Spectrum of activity
    1.   **Broad-spectrum** antimicrobials affect a wide range of bacteria.
    2.   Those that affect a narrow range are called **narrow-spectrum**.
  E.   Tissue distribution, metabolism and excretion of the drug
    1.   Some **antimicrobials cross the blood brain barrier** into the CSF; these can be used to treat meningitis.
    2.   **Drugs that are unstable in acid** cannot be taken orally and therefore must be administered through **injection**.
    3.   **Drugs that have a long half-life** need to be administered **less frequently**.
  F.   **Synergistic combinations** of drugs result in **enhanced antimicrobial activity**.
  G.   **Antagonistic antimicrobials interfere** with the activity of others.
  H.   Combinations are neither synergistic nor antagonistic are called **additive**.
  I.   Adverse effects include **allergies** to antimicrobials, **side effects** and **alteration of the normal flora**.
  J.   **Resistance** to antimicrobials is intrinsic or innate. Microorganisms can develop resistance through **spontaneous mutation** or the **acquisition of new genetic information**.

21.3   **Mechanisms of action** of antimicrobial drugs
    A.   **Antimicrobial drugs** target bacterial processes that utilize enzymes or structures that are either different, absent or not commonly found in eukaryotic cells.
    B.   Antibacterial medications that **inhibit cell wall synthesis**
        1.   The **β-lactam drugs** (**penicillins, cephalosporins, carbapenems** and **monobactams**) irreversibly **inhibit penicillin-binding proteins** (PBPs), ultimately leading to **cell lysis**. These drugs differ in their **spectrum of activity**.
        2.   **Vancomycin** blocks peptidoglycan synthesis.
        3.   **Bacitracin** interferes with the transport of peptidoglycan precursors.
    C.   Antibacterial medications **that inhibit protein synthesis**
        1.   The **prokaryotic 70S ribosome** serves as a target for selective toxicity.
        2.   Antibiotics include **aminoglycosides, tetracycline, macrolides, chloramphenicol, lincosamides, oxazolidinones** and **streptogramins**.
    D.   Antibacterial medications that **inhibit nucleic acid synthesis**
        1.   The **fluoroquinolones** interfere with DNA replication and transcription.
        2.   The **rifamycins** block initiation of transcription.
    E.   Antibacterial medications that **inhibit metabolic pathways**: **Sulfa drugs** and **Trimethoprim** inhibit enzymes.
    F.   Antibacterial medications that **interfere with cell membrane function**: Polymyxin B damages bacterial membranes.
    G.   Antibacterial medications that **interfere with processes essential to** *Mycobacterium tuberculosis*. Medications include **isoniazid, ethambutol** and **pyrazinamide.**

21.4   **Determining** the **sensitivity** of a bacterial strain to an antimicrobial drug
    A.   Determining the **minimum inhibitory concentration (MIC)** and **bactericidal concentrations (MBC)**
    B.   The **Kirby-Bauer disc diffusion test** is routinely used to qualitatively **determine the susceptibility** of a given organism to antimicrobial drugs.
    C.   **Antimicrobial susceptibility** can be determined by **automated methods** and the **E test** can be used to **determine the MIC.**

21.5   **Resistance** to antibacterial drugs
    A.   As antimicrobials are increasingly used and misused, the bacterial strains that are resistant to their effects have a selective advantage over their sensitive counterparts.
    B.   **Mechanisms** of resistance include **enzymes** that chemically modify a drug, **structural changes** in the target, **altered porin proteins** and **efflux pumps**.
    C.   **Vertical evolution** is the acquisition of resistance through **spontaneous mutation**.
    D.   **Horizontal evolution** is the acquisition of resistance through **gene transfer**.
    E.   The most common **mechanism of transfer of antibiotic resistance genes** is through the **conjugative transfer of R plasmids.**
    F.   The **emergence** and **spread of antimicrobial resistance can be slowed** by physicians **prescribing antimicrobials appropriately**, by patients carefully **following instructions** when taking antimicrobials and by **educating** the public about the appropriateness and limitations of antimicrobial therapy.

21.6   **Mechanism of action of antiviral drugs**
    A.   **Viruses use host cell machinery**, making them **difficult targets** for **selective toxicity**. There are **few antiviral drugs available** and they are generally **effective against only a specific type of virus**; none are able to eliminate latent viruses.
    B.   **Amantadine and rimantadine block** the **uncoating** of influenza A virus after it enters a cell.
    C.   **Nucleic acid synthesis: Nucleotide analogs interfere with replication** when they are incorporated into viral DNA.

D. **Protease inhibitors inhibit protease**, the enzyme required for the production of infectious HIV particles.

E. **Neuraminidase inhibitors** interfere with the release of influenza virus from a host cell.

21.7 **Mechanism of action of antifungal drugs: Few targets** for selectively toxic antifungal drugs.

A. Antifungal drugs that **inhibit plasma membrane synthesis and function** include **polyenes, azoles** and **allylamines**.

B. **Griseofulvin inhibits fungal cell division**.

C. **Flucytosine inhibits nucleic acid synthesis** and is used for **systemic yeast infections**.

21.8. Most **antiparasitic drugs** are thought to **interfere** with **biosynthetic pathways** of protozoan parasites or the neuromuscular function of worms.

Terms You Should Know

β-lactam ring
β-lactamase
Antagonistic
Antibiotic
Antimicrobial drug
Bacteriocidal
Bacteriostatic
Broad-spectrum antimicrobial
Chemotherapeutic agent
Ergosterol
Half-life

Horizontal evolution
Kirby-Bauer disc diffusion test
Minimum bactericidal concentration (MBC)
Minimum inhibitory concentration (MIC)
Narrow-spectrum antimicrobial
Neruaminidase inhibitor
Nucleoside analog
Penicillin
Penicillinase

Protease inhibitor
R plasmid
Selective toxicity
Semisynthetic
Sulfa drug
Synergistic
Therapeutic index
Vancomycin-resistant enterococci (VRE)
Vertical evolution
Zone of inhibition

Drugs References

Antibacterial Drugs

Penicillin G
Penicillin V
Methicillin
Dicloxicillin
Ampicillin
Ticarcillin
Piperacillin
Augmentin
Cephalosporins
   Cephalexin
   Cephradine
   Cefaclor
   Cefprozil
   Cefixime
   Cefibuten
   Cefepime
Carbapenems
Aztreonam
Vancomycin
Bacitracin

Aminoglycosides
   Streptomycin
   Gentamicin
   Tobramycin
   Amikacin
   Neomycin
Tetracyclines
   Tetracycline
   Doxycycline
Macrolides
   Erythromycin
   Clarithromycin
   Azithromycin
Chloramphenicol
Lincosamides
   Lincomycin
   Clindamycin
Oxazolidinones
   Linezolid

Streptogramins
   Quinupristin
   Dalfopristin
Fluoroquinolones
   Ciprofloxacin
   Ofloxacin
Rifamycins
   Rifampin
Sulfonamides
Trimethoprim
Polymyxin B
Ethambutol
Isoniazid
Pyrazinamide

Drugs References (continued)

Antiviral Drugs

Amantadine
Rimanatadine
Nucleoside analogs
 Acyclovir
 Ganciclovir
 Ribavirin
 Zidovudine (AZT)
 Dianosine (ddI)
 Lamivudine (3TC)
Nonnucleoside polymerase inhibitors
 Foscarnet
Nonnucleoside reverse transcriptase inhibitors
 Nevirapine
 Delavirdine
 Efavirenz
Protease inhibitors
 Indinavir
 Ritonavir
 Saquinavir
 Nelfinavir
Neuraminidase inhibitors
 Zanamivir
 Oseltamivir

Antifungal Drugs

Azoles
 Imidazoles
  Ketoconazole
  Miconazole
  Clotrimazole
 Triazoles
  Fluconazole
  Itraconazole
Polyenes
 Amphotericin B
 Nystatin
Allylamines
 Naftifine
 Tebinafine
Griseofulvin
Flucytosine

Anti-protozoan Drugs

Iodoquinol
Nitroimidazoles
 Metronidazole
Quinacrine
Folate antagonists
 Pyrimethamine
 Sulfonamide
Quinolones
 Chloroquine
 Mefloquine
 Primaquine
 Quinine
Eflornithine
Heavy metals
 Melarsoprol
 Sodium stibogluconate
 Meglumine antimonate
Nitrofuritimox

Anti-helminth Drugs

Avermectins
 Ivermectin
Benzimidazoles
 Mebendazole
 Thiabendazole
 Albendazole
Phenols
 Niclosamide
Piperazines
 Piperazine
 Diethylcarbamazine
Pyrazinoisoquinolines
 Praziquantel
Tetrahydropyrimidines
 Pyrantel pamoate
 Oxantel

Learning Activities

1. Identify the following:

First recognized successful antimicrobial agent	
First recognized antibiotic	

2. Define:

Term	Definition
Chemotherapeutic agent	
Antimicrobial drug or agent	
Antibiotic	
Selective toxicity	
Therapeutic index	
Bactericidal	
Bacteriostatic	
β-lactam drugs	

3. Differentiate between the following kinds of antimicrobials:

Broad-spectrum	
Narrow-spectrum	

4. List eight factors that one should consider in the selection of specific antimicrobial agent to treat an individual with a specific disease.

1.	
2.	
3.	
4.	
5.	
6.	
7.	
8.	

5. Define the following terms used to describe the use of combinations of antimicrobial drugs:

Term	Definition
Synergistic	
Antagonistic	
Additive	

6. List the three major types of adverse effects caused by antimicrobial agents.

1.	
2.	
3.	

7. Explain why antimicrobial agents are more effective against bacteria than against the eukaryotic pathogens and viruses.

8. Antibacterial drugs can be classified into groups by mode of action. List the six major modes of action of antibacterial drugs.

1.	
2.	
3.	
4.	
5.	
6.	

9. Match the following:

	1. Chloramphenicol	A.	Broad spectrum penicillin.
	2. Amphotericin B	B.	Can cause discoloration of the teeth.
	3. Cephalosporin	C.	Used to treat tuberculosis.
	4. Bacitracin	D.	Used to treat systemic fungal infections.
	5. Penicillin	E.	May cause aplastic anemia.
	6. Ethambutol	F.	Very toxic antifungal drug.
	7. Amoxicillin	G.	Very similar in structure and mode of action to penicillin.
	8. Vancomycin	H.	Used to treat malaria.
	9. Tetracyclines	I.	The first antibiotic discovered.
	10. Fluoroquinolones	J.	Only used topically to treat Gram-positive bacteria.
	11. Fluconazole	K.	Very toxic antibacterial drug.
	12. Chloroquine	L.	Broad-spectrum antibiotic used to treat urinary tract infections.

10. List the antibacterial agents that may be used to treat tuberculosis.

	Antibacterial agents
1.	
2.	
3.	
4.	

11. Give the uses and mode of action of the following antimicrobial agents used against bacterial infections.

Agent	Mode of action	Use
Penicillin G		
Penicillin V		
Tetracyclines		
Carboxypenicillins		
Cephalosporins		
Ampicillin		
Isoniazid		
Chloramphenicol		
Methicillin		
Bacitracin		
Aminopenicillins		
Streptomycin		
Vancomycin		
Ethambutol		
Neomycin		
Fluoroquinolones		
Erythromycin		
Polymyxin B		
Trimethoprim-sulfamethoxazole		

12. List four mechanisms of antimicrobial drug resistance.

1.	
2.	
3.	
4.	

13. Match the following antibiotics with their mode of action.

	Antibiotic	Mode of action
	1. Tetracyclines	A. Inhibits cell wall synthesis
	2. Chloramphenicol	B. Inhibits protein synthesis
	3. Ampicillin	C. Inhibits nucleic acid replication
	4. Sulfanilamide	D. Injures plasma membrane
	5. Neomycin	E. Inhibits synthesis of metabolites
	6. Erythromycin	
	7. Penicillins	
	8. Trimethoprim	
	9. Erythromycin	
	10. Fluoroquinolones	
	11. Polymyxin B	
	12. Cephalosporins	

14. Explain the action of the following enzymes.

β-lactamase	
Penicillinase	

15. Differentiate between vertical evolution and horizontal evolution in the acquisition of antimicrobial resistance.

16. Briefly describe or identify the following tests:

Test	Description
Determination of minimum inhibitory concentration (MIC)	
Determination of minimum bactericidal concentration (MBC)	
Diffusion assay	
Kirby-Bauer disc diffusion	
E test	

17. List the uses and modes of action of the following major antiviral agents used against viral diseases.

Agent	Mode of action	Use
Amantadine		
Acyclovir		
Ribavirin		
Ganciclovir		
Zidovudine (AZT)		
Foscarnet		
Nevirapine		
Indinavir		

18. List the uses and modes of action of the following antimicrobial agents used against fungal infections (antimycotic drugs).

Agent	Mode of action	Use
Amphotercin B		
Clotrimazole, Miconazole		
Ketoconazole		
Griesofulvin		
Naftifine		
Flucytosine		

19. List the uses and modes of action of the following major antiprotozoan drugs.

Agent	Mode of action	Use
Chloroquine		
Mefloquine		
Nitrofurtiox		
Metronidazole		

20. List the uses and modes of action of the following major antihelminth drugs.

Agent	Mode of action	Use
Niclosamide		
Praziquantel		
Mebendazole		
Pyantel pamoate		

Self Test

1. Which of the following antibiotics cannot be taken orally?

    a. penicillin G
    b. penicillin V
    c. ampicillin
    d. amoxicillin
    e. cephalosporin

2. Which of the following semisynthetic penicillins is resistant to penicillinase?

    a. penicillin G
    b. penicillin V
    c. methicillin
    d. amoxicillin
    e. piperacillin

3. Which of the following antibacterial agents is an alternative to penicillin?

    a. chloramphenicol
    b. ticarcillin
    c. neomycin
    d. cephalosporin
    e. isoniazid

4. Which of the following is used to treat bladder infections caused by *E. coli*?

    a. aztreonam
    b. vancomycin
    c. bacitracin
    d. dicloxicillin
    e. neomycin

5. Which of the following is an antibiotic used only for topical treatment?

    a. bacitracin
    b. vancomycin
    c. neomycin
    d. chloramphenicol
    e. Both a and c are correct.

6. Which of the following antibiotics can cause aplastic anemia?

    a. tetracylcline
    b. chloramphenicol
    c. fluoroquinolone
    d. sulfa drugs
    e. lincomycin

7. Which of the following antibiotics can be used to treat tuberculosis?

    a. isoniazid
    b. ethanbutol
    c. pyrazinamide
    d. rifampin
    e. All of the above.

8. Which of the following antiviral drugs can be used to treat influenza?

    a. amantadine
    b. acyclovir
    c. zidovudine
    d. nevirapine
    e. None of the above.

9. Which of the following antifungal drugs is best for the treatment of systemic infections?

    a. amphotericin B
    b. imidazole
    c. griseofulvin
    d. flucytosine
    e. None of the above.

10. Which of the following antimicrobial agents can be used to treat malaria?

    a. iodoquinol
    b. nitroimidazole
    c. quinolone
    d. nitrofurtimox
    e. praziquantel

Thought Questions

1. Why do we need to continue to develop new antimicrobial drugs?

2. Why can't we use the antibacterial drugs that inhibit cell metabolism to treat fungal, protozoan, helminthic and viral infections?

3. Explain why microbial resistance to antimicrobial agents is a major problem.

4. Explain the causes antimicrobial drug resistance.

Answers to Self Test Questions

1-a, 2-c, 3-d, 4-a, 5-e, 6-b, 7-e, 8-a, 9-a, 10-c

Chapter 22   Skin Infections

Overview

The skin is the largest and only visible organ of the human body. It is both a physical and chemical barrier that separates our internal environment from our external environment. While it is usually a means of protection, it can become infected either from external sources or from within. This chapter presents the major infectious diseases of the skin.

Learning Objectives

After studying the material in this chapter, you should be able to:

1. List the major groups of microorganisms making up the normal flora of the skin.
2. List the skin diseases caused by
    - *Staphylococcus aureus*
    - *Streptococcus pyogenes*
3. Give the causative agent for the following diseases:
    - Scalded skin syndrome
    - Streptococcal impetigo
    - Rocky Mountain spotted fever
    - Lyme disease
    - Chickenpox
    - Shingles
    - Measles
    - Rubella (German measles)
    - Warts
    - Superficial cutaneous mycoses
4. Give the major pathogenic and symptomatic features of the following:
    - Scalded skin syndrome
    - Streptococcal impetigo
    - Rocky Mountain spotted fever
    - Lyme disease
    - Chickenpox
    - Shingles
    - Measles
    - Rubella
5. Give the appropriate treatment or preventive measures for the following:
    - Streptococcal diseases
    - Chickenpox/Shingles
    - Measles
    - Rubella
    - Warts

6. Give the major routes of transmission of the following:
   - Streptococcal impetigo
   - Rocky Mountain spotted fever
   - Lyme disease
   - Chickenpox
   - Shingles
   - Measles
   - Rubella
   - Warts
7. Explain why rubella is considered such a significant disease.

Key Concepts

1. The skin is a large, complex organ that covers the external surface of the body.
2. The skin is a physical and chemical barrier to most microbial pathogens.
3. The normal flora of the skin help to protect against colonization by pathogens.
4. Normal flora can cause disease when body defense mechanisms are impaired.
5. Extensive skin damage can result from a toxin absorbed into the circulation from a localized infection.
6. Changes in the skin in an infectious disease reflect similar changes in other body tissues.

Summary Outline

22.1 The skin **repels potential pathogens** by shedding and being dry, acidic and toxic.
22.2 **Normal flora:** The skin is inhabited by large numbers of low virulence bacteria that help **prevent colonization** by more dangerous species.
   A. **Diphtheroids** are **Gram-positive, pleiomorphic, rod-shaped bacteria** that play a role in **acne** and **body odor. Fatty acids**, produced from the oily secretion of sebaceous glands, **keeps the skin acidic**.
   B. **Staphylococci** are **Gram-positive cocci** arranged **in clusters**. They are universally present and help **prevent colonization** by potential pathogens and **maintain the balance among flora** of the skin.
   C. Fungi: *Malassezia* sp. are **single-celled yeasts** found universally on the skin. They may cause **tinea versicolor**, dandruff, and serious skin disease in AIDS patients.
22.3 Bacterial skin diseases
   A. **Boils (furuncles)** and **carbuncles** are caused by *Staphylococcus aureus*, which is **coagulase-positive** and often **resists penicillin** and other antibiotics.
   B. **Staphylococcal scalded skin syndrome** results from **exotoxins** produced by certain strains of *Staphylococcus aureus*.
   C. **Impetigo** is a superficial skin infection caused by *Streptococcus pyogenes* and *Staphylococcus aureus*.
   D. **Acute glomerulonephritis** is caused by an **antibody-antigen reaction** and is an uncommon **complication of *S. pyogenes*** infections.
   E. **Rocky Mountain spotted fever** is caused by the **obligate intracellular bacterium** *Rickettsia rickettsii* and is a potentially fatal disease transmitted to humans by ticks.
   F. Lyme disease is caused by a spirochete, *Borrelia burgdorferi*, transmitted to humans by **ticks**.

22.4. **Skin diseases caused by viruses**
- A.   **Chickenpox (varicella)** is a common disease of childhood caused by the **varicella-zoster virus**, herpesvirus. **Shingles** or **herpes zoster**, can occur months or years after chickenpox, a reactivation of the varicella-zoster virus infection in the distribution of a sensory nerve. Shingles cases can be sources of chickenpox epidemics.
- B.   **Measles (rubeola)** is a potentially dangerous viral disease that can lead to serious secondary bacterial infections and fatal lung or brain damage. Measles can be controlled **by immunizing young children and susceptible adults with a live attenuated vaccine.**
- C.   **German measles (rubella)**, if contracted by a woman in the first eight weeks of pregnancy, results in a 90% chance of **birth defects** comprising the **congenital rubella syndrome. Immunization** with a **live attenuated virus** protects against this disease.
- D.   Other **viral rashes** of childhood
    1.   **Fifth disease (erythema infectiosum),** caused by **parvovirus B-19**, is characterized by a "slapped cheek" rash and can be fatal to people with certain anemias.
    2.   **Roseola (exanthem subitum)** is marked by several days of high fever and a transitory rash, which appears as the temperature returns to normal. It occurs mainly in infants six months to three years old. The disease is caused by **herpesvirus, type 6**.
- E.   **Warts** are **skin tumors** caused by a number of **papillomaviruses**. While they are generally benign, some **sexually transmitted papillomaviruses are associated with cancer of the uterine cervix**.

22.5   Skin diseases caused by fungi are **mycoses**. Invasive skin infections such as diaper rashes, may be caused by *Candida albicans*. Other fungi cause **athlete's foot, ringworm** and **invasions of the hair and nails**.

Terms You Should Know

Abscess	Exfoliatin	Mycosis
Capsule	Furuncle	Proteases
Carbuncle	H protein	Protein A
Coagulase	Hyaluronidase	Pyoderma
Dermatophyte	Intranuclear inclusion bodies	Sebum
Dermis	Keratin	Secondary infection
Diphtheroids	Koplik spots	Streptokinase
DNAse	Leukocidin	Streptolysins O and S
Epidermis	Lipase	Toxic shock syndrome toxin
Erythemia migrans	M protein	Zoonosis
Exanthem	Malaise	

Microorganisms to Know

Borrelia burgdorferi	*Malassezia* species	*Staphylococcus aureus*
Candida albicans	*(Pityrosporum)*	*Staphylococcus epidermidis*
Corynebacterium diphtheriae	*Microsporum*	*Streptococcus pneumoniae*
Dermacentor andersoni	*Papillomaviruses*	*Streptococcus pyogenes*
Dermacentor variabilis	*Propionibacterium acnes*	*Trichophyton*
Epidermophyton	*Rickettsia rickettsii*	*Varicella-zoster virus*
Haemophilus influenzae	*Rubella virus*	
Ixodes scapularis	*Rubeola virus*	

Diseases to Know

Acne
Acute glomerulonephritis
Chickenpox (varicella)
Congenital rubella syndrome
Congenital varicella syndrome
Fifth disease (erythema infectiosum)
Folliculitis

German measles (rubella)
Lyme disease
Measles (rubeola)
Plantar warts
Reye's syndrome
Rocky Mountain spotted fever
Roseola (exanthem subitum, roseola infantum)

Scalded skin syndrome
Shingles (Herpes-zoster)
Streptococcal impetigo
Subacute sclerosing panencephalitis
Superficial cutaneous mycoses
Tinea versicolor
Warts

Learning Activities

1. List the major groups of normal flora of the skin.

1.	
2.	
3.	

2. Match the disease with its causative agent:

Disease	Causative agent
1. Acne	A. Togavirus
2. Warts	B. Poxvirus
3. Chickenpox	C. *Borrelia burgdorferi*
4. Shingles	D. *Malassezia* species
5. Measles	E. Herpes simplex
6. German measles	F. Herpesvirus type 6
7. Rocky Mountain spotted fever	G. *Propionibacterium acnes*
8. Lyme disease	H. Varicella-Zoster virus
9. Roseola	I. *Rickettsia rickettsii*
10. Rubella	J. Paramyxovirus
11. Fifth disease	K. Human parvovirus
12. Superficial cutaneous mycosis	L. Papovaviruses (Papillomaviruses)

3. Match the following diseases with their causative agent:

	Disease	Causative agent
	1. Scalded skin syndrome	A. *Staphylococcus aureus*
	2. Impetigo	B. *Streptococcus*
	3. Toxic shock syndrome	C. Both organisms
	4. Folliculitis	D. Neither organism
	5. Furuncle	
	6. Fifth disease	

4. Give the major diagnostic features of the following:

Disease	Major diagnostic features
Impetigo	
Scalded skin syndrome	
Rocky Mountain Spotted fever	
Chickenpox	
Shingles	
Measles	
Rubella	

5. Give the major routes of transmission for the following diseases:

Disease	Route of transmission
Impetigo	
Warts	
Chickenpox	
Shingles	
Measles	
Rubella	

6. Give the appropriate treatment or preventive measure for the following:

Disease	Treatment or prevention
Streptococcal diseases	
Scalded skin syndrome	
Rocky Mountain spotted fever	
Lyme disease	
Chickenpox/Shingles	
Measles	
Rubella	
Warts	
Superficial cutaneous mycoses	

7. Why do we vaccinate to prevent rubella?

Self Test

1. The outermost layer of the skin is the

    a. dermis.
    b. epidermis.
    c. hypodermis.
    d. keratin.
    e. sebum.

2. The protein found in the skin that forms a significant barrier to microbes is

    a. sebum.
    b. keratin.
    c. coagulase.
    d. M-protein.
    e. protein A.

3. Normal flora of the skin includes

    a. *Corynebacterium diphtheriae.*
    b. *Propionibacterium acnes.*
    c. *Staphylococcus species.*
    d. *Malassezia* species (*Pityrosporum*).
    e. All of the above.

4. *Staphylococcus aureus* may be the causative agent of all of the above EXCEPT:

    a. furuncles
    b. scalded skin syndrome
    c. impetigo
    d. tinea versicolor
    e. carbuncles

5. Lyme disease is caused by

    a. *Propionibacterium acnes.*
    b. *Streptococcus pyogenes.*
    c. *Rickettsia rickettsii.*
    d. *Borrelia burgdorferi.*
    e. *Epidermophyton.*

6. Varicella-zoster virus causes

    a. measles.
    b. German measles.
    c. chickenpox.
    d. Fifth disease.
    e. English measles.

7. Measles and German measles are similar in that

    a. they both are caused by a Poxvirus.
    b. they both cause congenital infections resulting in birth defects.
    c. they both are transmitted by the respiratory route.
    d. they both are mild diseases in children.
    e. None of the above are true.

8. Fifth disease

    a. is also known as exanthem subitum.
    b. is characterized by a "slapped cheek" rash.
    c. is caused by herpes virus type 6.
    d. All three statements are true.
    e. Only b and c are true.

9. Warts are caused by

    a. a poxvirus.
    b. a togavirus.
    c. a papillomavirus.
    d. a parvovirus.
    e. a herpesvirus.

10. A tick is the vector for which of the following diseases?

    a. Lyme disease
    b. Roseola
    c. Rocky Mountain spotted fever
    d. All of the above.
    e. Both a and c.

Thought Questions

1. How does the normal flora of the skin help to protect against infection?

2. Why is *Staphylococcus aureus* a major pathogen, while other species of *Staphylococcus* make up part of the normal flora?

Answers to Self Test Questions

1-b, 2-b, 3-e, 4-d, 5-d, 6-c, 7-c, 8-e, 9-c, 10-e

Chapter 23   Respiratory System Infections

Overview

Respiratory tract infections are the most common kind of human infections. Upper respiratory tract infections are usually not-life threatening, but nevertheless produce an unpleasant, inconvenient experience. Some can lead to serious complications. Lower respiratory tract infections are usually more serious including various types of pneumonia, a common cause of death. In this chapter the infections of the upper and lower respiratory tract are presented.

Learning Objectives

After studying the material in this chapter, you should be able to:

1. Describe the components and distribution of normal flora of the respiratory system.
2. List the cause, major characteristics and modes of transmission for the following diseases:
    *   Streptococcal pharyngitis
    *   Scarlet fever
    *   Diphtheria
    *   Conjunctivitis (pinkeye)
    *   Otitis media
    *   Sinus infections
    *   Common cold
    *   Adenoviral phayrngitis
    *   Pneumococcal pneumonia
    *   Klebsiella pneumonia
    *   Mycoplasmal pneumonia
    *   Whooping cough
    *   Tuberculosis
    *   Legionellosis (Legionnaires' disease)
    *   Influenza
    *   Respiratory syncytial virus infections
    *   Hantavirus pulmonary syndrome
    *   Coccidioidomycosis (valley fever)
    *   Histoplasmosis
3. Name the reservoirs for infectious microorganisms that cause the following diseases:
    *   Tuberculosis
    *   Histoplasmosis
    *   Influenza
4. Give the virulence factors for the following diseases:
    *   Diphtheria
    *   Scarlet fever
    *   Pneumococcal pneumonia
    *   Klebsiella pneumonia
    *   Whooping cough

5. Explain why epidemics of influenza recur.
6. Define of identify the Guillian-Barré syndrome.
7. List the bacteria that cause pneumonia and give the important features of each.
8. Describe the following organisms:
   - *Corynebacterium*
   - *Mycobacterium*
   - *Mycoplasma*
   - *Bordetella*
   - *Klebsiella*
9. List the types of vaccines associated with the following diseases:
   - Tuberculosis
   - Whooping cough
   - Diphtheria
   - Influenza

Key Concepts

1. The nasal cavity, nasopharynx and pharynx are colonized by normal flora.
2. The mastoid sinuses, middle ear cavity, nasal sinuses, trachea, bronchi, bronchioles and alveoli are sterile under normal conditions.
3. Pharyngitis is a common problem caused by a number of different bacterial species; most do not require antibiotic therapy.
4. Untreated infections of the skin or throat caused by *Streptococcus pyogenes* can result in serious heart or kidney problems due to antigen-antibody complexes.
5. Some diseases are toxin-mediated with the toxin causing the damage to specific sites remote from the site of infection.
6. The causes of viral infections of the upper respiratory tract are usually not determined because there are many infectious agents that produce the same signs and symptoms of disease.
7. Because of efficient defense mechanisms, bacterial infections of the lungs are usually limited to individuals with impaired immunity.
8. The most important infection of the lower respiratory tract is pneumonia, which is usually caused by bacteria or viruses, but may be caused by eukaryotic organisms, allergies or chemicals.
9. Vaccines may reduce the incidence of some lower respiratory tract infections such as pertussis and some types of pneumonia.
10. Tuberculosis infections are often difficult to treat and may result in latency leading to reactivation later in life.
11. Viral infections like influenza are very common and widespread resulting in a self-limited, but unpleasant disease.
12. Deaths from diseases like influenza are usually due to secondary bacterial infections.

Summary Outline

23.1 Structure and Function
   A. The respiratory system is lined with mucous membranes.
   B. The function of the respiratory system is (1) temperature and humidity regulation of inspired air, (2) removal of microorganisms and debris and (3) exchange of gases between the blood and the external environment.

C. . The ciliated cells that line the respiratory tract remove microorganisms by a constant sweeping action.

23.2 **Normal flora** of the nasal cavity includes **diphtheroids**, and *Staphylococcus aureus*, coagulase-positive staphylococci. Viruses and microorganisms are normally absent from the lower respiratory system.

23.3 Bacterial infections of the upper respiratory system

A. *Streptococcus pyogenes* causes strep throat (**streptococcal pharyngitis**), a significant bacterial infection that may lead to **scarlet fever, rheumatic fever, toxic shock** or **glomerulonephritis**.

B. **Diphtheria**, caused by *Corynebacterium diphtheriae*, is a **toxin-mediated** disease that can be prevented by **immunization**.

C. **Conjunctivitis** (pink eye) is usually caused by *Haemophilus influenzae* or *Streptococcus pneumoniae*, the pneumococcus. Viral causes, including **adenoviruses** and **rhinoviruses**, usually result in a milder illness.

D. **Otitis media** and **sinusitis** develop when infection extends from the nasopharynx.

23.4 **Viral infections of the upper respiratory system**

A. The **common cold** can be caused by many different viruses, **rhinoviruses** being the most common.

B. **Adenoviruses** cause illnesses varying from mild to severe, which can resemble a common cold or strep throat.

23.5 **Bacterial infections of the lower respiratory system**

A. *Streptococcus pneumoniae,* the cause of pneumococcal pneumonia, is virulent because of its capsule.

B. *Klebsiella pneumoniae*, Gram-negative rod-shaped bacteria, causes a pneumonia that is representative of many **nosocomial pneumonias** that cause **permanent damage to the lung** such as abscesses. Treatment is more difficult, partly because *Klebsiella* often contains R factor plasmids.

C. **Mycoplasmal pneumonia** is often called walking pneumonia; serious complications are rare. **Penicillins** and **cephalosporins are not useful** in treatment because the cause, *M. pneumoniae*, **lacks a cell wall**.

D. **Whooping cough (pertussis)** is characterized by violent spasms of coughing and gasping and is caused by the Gram-negative rod, *Bordetella pertussis*. Childhood **immunization** prevents the disease.

E. **Tuberculosis**, caused by the **acid-fast rod** *Mycobacterium tuberculosis*, is slowly progressive or heals and remains latent, presenting the risk of later reactivation.

F. **Legionnaires' disease** occurs when there is a high infecting dose of *Legionella pneumophila*, a rod-shaped bacterium common in the environment.

23.6 **Viral infections of the lower respiratory system**

A. **Influenza**: Widespread epidemics are characteristic of **influenza A viruses**. **Antigenic shifts** and **drifts** are responsible. Deaths are usually caused by secondary infection. **Reye's syndrome** may rarely occur during recovery from **influenza B** but is probably not caused by the virus itself.

B. **Respiratory syncytial virus (RSV)** is the leading cause of serious respiratory disease in **infants** and **young children**.

C. **Hantavirus pulmonary syndrome** is contracted from inhalation of dust infected by mice with the virus and is often fatal.

23.7 Fungal infections of the lung

A. **Coccidioidomycosis (Valley fever)** occurs in hot, dry areas of the Western Hemisphere and is initiated by airborne spores of the dimorphic soil fungus *Coccidioides immitis*.

B. **Histoplasmosis (Spelunker's disease)** occurs in tropical and temperate zones around the world. The causative fungus, ***Histoplasma capsulatum***, is dimorphic and found in soils contaminated by bat or bird droppings.

Terms You Should Know

Antigenic drift
Antigenic shift
Bronchiolitis
Bronchitis
Caseous necrosis
Conjunctivitis
Dacryocystitis
Epiglottis

External otitis
Goblet cells
Granuloma
Laryngitis
Mastoiditis
Mucociliary escalator
Otitis media
Paroxysmal coughing

Pharyngitis
Pleurisy
Pneumonia
Pneumonitis
Rhinitis
Sinusitis
Sputum
Tonsillitis

Microorganisms to Know

β-hemolytic streptococci
    Lancefield group C
Adenoviruses
Bacteroides
Bordetella pertussis
Coccidioides immitis
Corynebacterium diphtheriae
Haemophilus influenzae

Hantaviruses
Histoplasma capsulatum
Influenza A virus
Klebsiella pneumoniae
Legionella pneumophila
Moraxella lacunata
Mycobacterium bovis
Mycobacterium tuberculosis

Mycoplasma pneumoniae
Neisseria gonorrhoeae
Respiratory syncytial virus
Rhinoviruses
Staphylococcus aureus
Streptococcus pyogenes
Streptococcus pneumoniae

Diseases to Know

Acute rheumatic fever
Adenoviral pharyngitis
Chorea
Coccidioidomycosis (valley
    fever)
Common cold
Diphtheria
Earache
Guillain-Barré syndrome

Hantavirus pulmonary syndrome
Histoplasmosis (spelunkers'
    disease)
Influenza
Klebsiella pneumonia
Legionnaires' disease
Mycoplasmal pneumonia
Pinkeye
Pneumococcal pneumonia

Quinsy
Respiratory syncytial virus
    infections
Scarlet fever
Sinus infections
Streptococcal pharyngitis (Strep
    throat)
Tuberculosis
Whooping cough (Pertussis)

Learning Activities

1. List the normal flora of the respiratory system.

1.	
2.	
3.	
4.	
5.	
6.	

2. Give the cause, major characteristics, and mode of transmission for the following:

Disease	Cause	Major characteristics	Mode of transmission
Streptococcal pharyngitis			
Scarlet fever			
Diphtheria			
Otitis media			
Common cold			
Whooping cough			
Tuberculosis			
Legionellosis			

3. Name the reservoirs for the infectious agents that cause the following diseases:

Disease	Reservoir
Tuberculosis	
Histoplasmosis	
Influenza	

4. List the virulence factors for the following diseases:

Disease	Virulence factors
Diphtheria	
Scarlet fever	
Pneumococcal pneumonia	
Klebsiella pneumonia	
Whooping Cough	

5. On the drawing of components of the cell envelope of *Streptococcus pyogenes* label the following:

    M protein, Protein G, Protein F, Hyaluronic acid capsule, Peptidoglycan, Cytoplasmic membrane, Group carbohydrate, Lipoteichoic acid

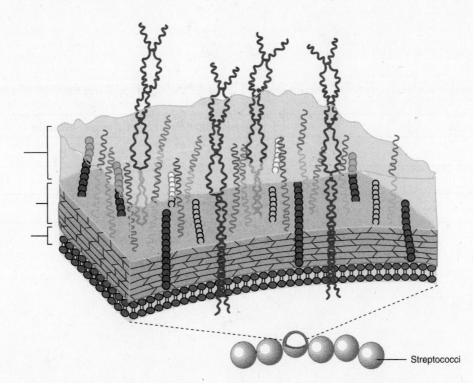

Streptococci

6. On the drawing of the influenza virus label the following:

Neuraminidase spike, Hemagglutinin spike, Lipid envelope, RNA, Matrix protein

7. Explain why epidemics of influenza recur.

8. List the major bacterial causes of pneumonia and describe the distinguishing features of each.

	Bacteria	Important features of the pneumonia
1.		
2.		
3.		

9. Indicate the major features of the following organisms:

Corynebacterium	
Mycobacterium	
Mycoplasma	
Bordetella	
Klebsiella	

10. List the types of vaccines associated with the following diseases:

Tuberculosis	
Whooping cough	
Diphtheria	
Influenza	

Self Test

1. Which of the following may occur following recovery from streptococcal pharyngitis?

    a. scarlet fever
    b. acute glomerulonephritis
    c. rheumatic fever
    d. Both a and b are correct.
    e. Both b and c are correct.

2. The rash of scarlet fever is caused by

    a. an erythrogenic toxin.
    b. bacteria in the skin.
    c. high fever.
    d. a virus.
    e. an allergy.

3. Acute glomerulonephritis, a consequence of streptococcal throat infections, is actually caused by

    a. immune complexes in the kidneys.
    b. *Streptococcus pyogenes* endotoxins.
    c. immune complexes in the throat.
    d. *Streptococcus pyogenes* exotoxins.
    e. Both a and d are correct.

4. Which of the following is an important cause of otitis media (middle ear infection)?

    a. *Streptococcus pyogenes*
    b. *Streptococcus pneumoniae*
    c. *Mycoplasma pneumoniae*
    d. *Haemophilus influenzae*
    e. Both b and d are correct.

5. Which of the following causes walking pneumonia?

    a. *Mycoplasma tuberculosis*
    b. *Mycobacterium pneumoniae*
    c. *Klebsiella pneumoniae*
    d. *Mycobacterium tuberculosis*
    e. *Mycoplasma pneumoniae*

6. Histoplasmosis is associated with the growth of *Histoplasma capsulatum* in

    a. soil containing bird and bat droppings.
    b. dust and dry air.
    c. plants.
    d. hospitals.
    e. nasal secretions.

7. Deaths from influenza are most often due to

    a. heart failure.
    b. secondary bacterial infections.
    c. exotoxins.
    d. kidney failure.
    e. suffocation.

8. *Mycoplasma pneumoniae* infections are most common in

    a. people over the age of 40.
    b. infants.
    c. adolescents.
    d. people over the age of 60.
    e. people between 40 and 60.

9. Person-to-person transmission has not been demonstrated for which of the following bacteria?

    a. *Legionella pneumophila*
    b. *Mycoplasma pneumoniae*
    c. *Klebsiella pneumoniae*
    d. *Mycobacterium tuberculosis*
    e. All of the above may be transmitted by person-to-person contact.

10. The part of the DPT vaccine that produces immunity to whooping cough is

    a. attenuated *Bordetella pertussis*.
    b. a part of the *Bordetella pertussis* bacterial cell.
    c. killed *Bordetella pertussis*.
    d. pertussis endotoxin.
    e. pertussis toxoid.

Thought Questions

1. Why has it been so difficult to make a vaccine to prevent the common cold and to produce an effective treatment for the common cold?

2. Why do epidemics of the "flu" continue to occur although we have an effective vaccine?

3. Why is it a bad idea to take antibiotics when you have a cold?

Answers to Self Test Questions

1-e, 2-a, 3-a, 4-e, 5-e, 6-a, 7-b, 8-c, 9-a, 10-b

Chapter 24   Alimentary System Infections

Overview

The alimentary system is the second most common target for human infections. Diarrheal disease results in a significant mortality rate worldwide. While a large variety of microorganisms enter the body through food, the fecal-oral route of transmission is the principal means of spread of pathogens. The stomach and its acid contents constitute an effective barrier to many microorganisms, but not to all. Many microorganisms have adapted means to get through the stomach unharmed leading to significant infections in the lower alimentary tract. This chapter presents the major diseases of the alimentary system.

Learning Objectives

After studying the material in this chapter, you should be able to:

1. Describe the pathogenesis of dental caries.
2. Describe the role of *Helicobacter pylori* in the pathogenesis of gastric and duodenal ulcers.
3. Describe the symptoms, causes, modes of transmission and treatment and/or prevention of the following diseases:
   - Herpes simplex
   - Mumps
4. List the characteristics, causes, modes of transmission, prevention and treatment for the following diseases:
   - Cholera
   - Shigellosis
   - *Escherichia coli* gastroenteritis
   - Salmonellosis
   - Typhoid fever
   - Campylobacteriosis
5. List the symptoms, causes, pathogenesis, modes of transmission, treatment, if any, and prevention of the following:
   - Rotaviral gastroenteritis
   - Norwalk virus gastroenteritis
   - Hepatitis A
   - Hepatitis B
   - Hepatitis C
6. List the symptoms, cause, pathogenesis, modes of transmission, treatment, if any, and prevention of the following protozoan diseases:
   - Giardiasis
   - Cryptosporidiosis
   - Cyclosporiasis
   - Amebiasis

Key Concepts

1. The alimentary tract is a major portal of entry for pathogenic microorganisms.
2. Infectious diseases of the alimentary tract are very common and widespread.
3. Worldwide, the effects of diarrhea are often deadly.
4. Normal flora of the alimentary tract, especially in the mouth and large intestine, help to prevent many infections.
5. The pathogenesis of dental caries depends on diet and presence of acid-forming bacteria.
6. Infections of the esophagus are unusual and their occurrence suggests immunodeficiency.
7. The acidic environment of the stomach constitutes an efficient barrier to microorganisms, but some have adaptations that allow them to escape destruction there.
8. Infections of the stomach were once considered unusual, but are now considered common.
9. Some bacterial pathogens activate a type III secretion system when they make contact with a host cell. This results in the joining of the cytoplasm of the bacterial and host cells by a short flagellum allowing gene products to pass directly into the host cell to activate endocytosis or other processes that benefit the bacterium.
10. Microbial toxins can alter the secretory activity of cells of the intestinal lining without killing or visibly damaging them, but can still cause disease, especially diarrhea.
11. Some alimentary tract infections produce a carrier state, which may result in a source of infection that lasts for many years.
12. Food poisoning is very common and may result from either the ingestion of infectious microorganisms with food, or microbial toxins ingested with food.

Summary Outline

25.1 The alimentary tract is a tube from the mouth to the anus and comprises a major portal of entry for pathogens.

24.2 Normal flora

    A.   The **mouth**: The species of bacteria that inhabit the mouth **vary with location**. **Dental plaque** contains bacteria attached to teeth or each other. Teeth allow for colonization by **anaerobic bacteria**.

    B.   The **intestine**: While the stomach is normally devoid of microorganisms, about one-third of the weight of feces is due to microorganisms, mostly **anaerobes**. The biochemical activities of microorganisms in the large intestine include **synthesis of vitamins**, **degradation of indigestible substances**, **competitive inhibition of pathogens**, **production of cholesterol**, **chemical alteration of medications** and **production of carcinogens**.

24.3 **Bacterial diseases of the upper alimentary system**

    A.   **Dental caries** is caused mainly by *Streptococcus mutans* involved in formation of extracellular **glucans from dietary sucrose**. Penetration of the calcium phosphate tooth structure depends on **acid production** by cariogenic dental plaque. *S. mutans* is **not inhibited by acid** and stores fermentable intracellular polysaccharide. **Control of dental caries** depends mainly on supplying **fluoride** and **restricting dietary sucrose.**

    B.   **Periodontal disease** is caused by an inflammatory response to the plaque bacteria at the gum line.

    C.   **Trench mouth**, or **acute necrotizing ulcerative gingivitis (ANUG)**, can occur at any age in association with **poor mouth care**.

    D.   *Helicobacter pylori* predisposes the stomach and the uppermost part of the duodenum to **peptic ulcers**. Treatment with antimicrobial medications can cure the infection and prevent peptic ulcer recurrence.

24.4 Viral diseases of the upper alimentary system

    A.   **Herpes simplex**, caused by an **enveloped DNA virus**, usually begins in the mouth and throat; esophageal infections suggest immunodeficiency. HSV persists as a **latent infection** inside sensory nerves; production of active disease occurs when the body is stressed.

    B.   **Mumps is caused by an enveloped** RNA virus that infects not only the **parotid glands**, but also the **meninges**, **testicles** and other body tissues. Mumps virus generally causes more severe disease in persons beyond the age of puberty; it can be **prevented using a live attenuated vaccine**.

24.5 Bacterial diseases of the lower alimentary tract

    A.   **Cholera** is a severe form of diarrhea caused by a **toxin of *Vibrio cholerae*** that acts on the small intestinal epithelium.

    B.   **Shigellosis** is caused by species of *Shigella*, common causes of **dysentery** because they **invade the epithelium** of the large intestine.

    C.   *Escherichia coli* gastroenteritis: Virulence often depends on **plasmids**. Some strains, such as **O 157 B 7**, can cause **hemolytic uremic syndrome**.

    D.   **Salmonellosis**: Caused by strains of *Salmonella*, which is often **food-borne** and commonly found in **eggs** and **poultry**.

    E.   **Typhoid fever,** which only infects humans, is caused by *Salmonella typhus* and is characterized by **high fever**, **headache** and **abdominal pain. Untreated**, it has a **high morality rate**. An **oral attenuated vaccine** helps prevent the disease.

    F.   **Campylobacteriosis**, caused by *Campylobacter jejuni,* is the **most common bacterial cause of diarrhea** in the United States.

24.6 Viral diseases of the lower alimentary tract

    A.   **Rotaviral gastroenteritis** is the main diarrheal illness of infants and young children, but can involve adults, as in **traveler's diarrhea**.

    B.   **Norwalk virus gastroenteritis** accounts for almost half the cases of gastroenteritis in the United States.

    C.   **Hepatitis A** is usually **mild or asymptomatic in children**; some cases are prolonged, with **weakness, fatigue and jaundice**. Hepatitis A virus (HAV) is a **picornavirus spread by fecal contamination** of hands, food or water. An injection of **gamma globulin gives temporary protection** from the disease. An **inactivated vaccine** is available to immunize against the disease.

    D.   **Hepatitis B (HBV)** is a hepadnavirus **spread by blood, blood products, semen and from mother to baby**; it is generally more severe than hepatitis A. **Asymptomatic carriers** are common and can unknowingly transmit the disease. **Chronic infection** is common and can lead to scarring of the liver **(cirrhosis)** and **liver cancer**.

    E.   **Hepatitis C virus (HCV)** is a flavivirus transmitted mainly by blood; 60% of cases may be acquired from **needle sharing** by injecting drug abusers. **Hepatitis C is asymptomatic** in over 60% of acute infections; 80% of infections become **chronic**.

24.7 Protozoan diseases of the lower alimentary tract

    A.   **Giardiasis** is caused by *Giardia lamblia*, a Mastigophoran, and is usually transmitted by **drinking water contaminated by feces**. It is a common cause of **traveler's diarrhea**.

    B.   **Cryptosporidiosis** is caused by *Cryptosporidium parvum,* a member of the Apicomplexa. The **oocysts** are infectious, resist chlorination and are too small to be removed by most filters.

C.   **Cyclosporiasis** is cause by *Cyclospora cayetanensis*, which is transmitted by the fecal-oral, via water or produce such as berries; it causes **traveler's diarrhea**. There is no person-to-person spread; no hosts other than humans are known.

D.   **Amebiasis** is caused by *Entamoeba histolytica*; it is an important cause of **dysentery**; **often chronic**; infection can spread to the liver and other organs.

Terms You Should Know

Amebic dysentery	Gastritis	Herpetic whitlow
Dental calcullus	Gastroenteritis	Infectious hepatitis
Dental caries	Gingival crevice	Microvilli
Dental plaque	Gingivitis	Pepsin
Dysentery	Hemolytic uremic syndrome	Serum hepatitis
Enteric fever	Herpes simplex labialis	Shiga toxin

Microorganisms to Know

Campylobacter jejuni	Hepatitis A virus (HAV)	*Salmonella* species
Cryptosporidium parvum	Hepatitis B virus (HBV)	*Salmonella typhi*
Cyclospora cayetanensis	Hepatitis C virus (HCV)	*Shigella* species
Entamoeba histolytica	Herpes simplex (HSV)	*Streptococcus mutans*
Escherichia coli	Norwalk virus	*Treponema* species
Fusobacterium species	*Porphyromonas gingivalis*	*Vibrio cholerae*
Giardia lamblia	*Prevotella* species	
Helicobacter pylori	Rotavirus	

Diseases to Know

Amebiasis	Giardiasis	Norwalk virus gastroenteritis
Campylobacteriosis	*Helicobacter pylori* gastritis	Periodontal disease
Cholera	Hepatitis A	Rotaviral gastroenteritis
Cryptosporidiosis	Hepatitis B	Salmonellosis
Cyclosporiasis	Hepatitis C	Shigellosis
Dental caries	Herpes simplex	Trench mouth
Escherichia coli gastroenteritis	Mumps	Typhoid fever

Learning Activities

1. Compare and contrast the pathology of dental caries and periodontal disease.

2. Give the characteristics, cause, mode of transmission, prevention and treatment for the
    following diseases:

Disease	Characteristics	Cause	Transmission	Prevention and treatment
Cholera				
Shigellosis				
Escherichia coli gastroenteritis				
Salmonellosis				
Typhoid fever				
Campylobacteriosis				

3. Give the characteristics, cause, transmission, treatment, if any, and prevention of the following:

	Characteristics	Cause	Transmission	Prevention and treatment
Giardiasis				
Cryptosporidiosis				
Cyclosporiasis				
Amebiasis				

4. Give the characteristics, cause, transmission, treatment, if any, and prevention of the following:

	Characteristics	Cause	Transmission	Prevention and treatment
Herpes simplex				
Mumps				
Rotaviral gastroenteritis				
Norwalk virus gastroenteritis				
Hepatitis A				
Hepatitis B				
Hepatitis C				

5. Compare and contrast hepatitis A, hepatitis B, and hepatitis C as to type of virus, disease caused, prevention and/or treatment.

Self Test

1. Herpes simplex virus can remain latent in a noninfectious form within

    a. the lips.
    b. the mouth.
    c. saliva.
    d. sensory nerves.
    e. None of the above.

2. *Helicobacter pylori* infections are associated with which of the following?

    a. gastritis
    b. peptic ulcers
    c. duodenal ulcers
    d. gastric cancer
    e. All of the above.

3. Which of the following is commonly associated with recurrences of herpes simplex virus infections?

    a. fever
    b. sunburn
    c. menstruation
    d. All of the above.
    e. Only a and b are correct.

4. The severe swelling and pain associated with mumps is due to

    a. endotoxins.
    b. an inflammatory response.
    c. immune complexes.
    d. exotoxins.
    e. All of the above are correct.

5. Which of the following are often associated with eating chicken?

    a. *Escherichia coli* gastroenteritis
    b. Salmonellosis
    c. Giardiasis
    d. Campylobacteriosis
    e. Both b and d are correct.

6. Which of the following type(s) of hepatitis is primarily transmitted by the fecal-oral route?

    a. Hepatitis A
    b. Hepatitis B
    c. Hepatitis C
    d. Hepatitis E
    e. Both a and d are correct.

7. The most common cause of post-transfusion hepatitis is

    a. hepatitis A virus.
    b. hepatitis B virus.
    c. hepatitis C virus.
    d. Both a and b are correct.
    e. All three are correct.

8. Which of the following is responsible for most of the waterborne disease epidemics in the United States in which a cause is identified?

    a. *Salmonella*
    b. *Escherichia coli*
    c. *Shigella*
    d. *Campylobacter jejuni*
    e. *Giardia lamblia*

9. To which of the following groups of *Escherichia coli* does the O 157 E strain belong?

    a. Enterotoxigenic
    b. Enteroinvasive
    c. Enteropathogenic
    d. Enterohemorrhagic
    e. None of these.

10. Vaccines have been developed for which of the hepatitis viruses?

    a. Hepatitis A (HAV)
    b. Hepatitis B (HBV)
    c. Hepatitis C (HCV)
    d. Both a and b are correct.
    e. Both b and c are correct.

Thought Questions

1. Describe a public health plan to limit the spread of viral hepatitis B.

2. Diarrheal disease is a major health problem worldwide, resulting in the death of many people. Develop a plan to intervene and treat this problem.

Answers to Self Test Questions

1-d, 2-e, 3-d, 4-b, 5-b, 6-a, 7-c, 8-e, 9-d, 10-d

Chapter 25   Genitourinary Infections

Overview

The urinary tract is one of the most common sites for opportunistic infections, originating from the normal flora when defense mechanisms are weakened. Anything that interferes with the normal flow of urine increases the risk of infection. Females are more susceptible to urinary tract infections for anatomical reasons. Most genital tract infections are sexually transmitted, but some are not. Sexually transmitted diseases (STDs) including gonorrhea, genital herpes and AIDS present a growing worldwide problem. This chapter presents the infections of the genitourinary tract.

Learning Objectives

After studying the material in this chapter, you should be able to:

1. Describe the defense mechanisms that help to prevent infections in the urinary tract.
2. Explain why nosocomial infections often involve the urinary tract.
3. Describe the diagnosis, prevention and treatment of urinary tract infections.
4. List the characteristics, site of the disease, causes, modes of transmission, prevention and treatment for the following diseases:
    - Cystitis
    - Pyelonephritis
    - Leptospirosis
    - Bacterial vaginosis
    - Vulvovaginal candidiasis
    - Toxic shock syndrome
5. Differentiate between non-sexually transmitted diseases of the genital tract and sexually transmitted (STDs) of the genital tract.
6. Explain why sexually transmitted diseases (STDs) are considered a major problem and how they can be controlled within a population.
7. Describe the prevalence, characteristics, cause, modes of transmission, treatment and complications of gonorrhea.
8. Describe the prevalence, characteristics, cause, modes of transmission, treatment and complications of syphilis.
9. List the characteristics, causes, transmission, treatment, if any, and prevention of the following.
    - Chlamydial genital system infections
    - Chancroid
    - Genital herpes simplex
    - Genital warts
    - AIDS

Key Concepts

1. Under normal circumstances the urine and the urinary tract are free of microorganisms above the entrance of the bladder.
2. The washing action of the urinary flow plays a major role in prevention of bladder infections.
3. Hormone levels, especially estrogens, play a significant role in the balance of normal flora in the vagina. This results in varying resistance to infections.
4. Genital tract infections usually require direct person-to-person transmission. Some of these are sexually transmitted and some are not.
5. Sexually transmitted diseases may or may not be symptomatic.
6. Both symptomatic and asymptomatic infections can cause significant damage to the genital tract and can be spread to others.
7. The chance of acquiring a sexually transmitted disease increases with the number of sexual partners.
8. Simple preventive measures are highly successful in preventing sexually transmitted diseases.
9. Viral STDs are less common than bacterial STDs, but no cures are currently known.
10. Gonorrhea, one of the most prevalent sexually transmitted diseases, is slowly becoming more resistant to antimicrobial therapy.
11. In spite of advances in treatment, the worldwide death rate due to STDs continues to be in the millions each year because of the lack of access to therapy, and the lack of an effective vaccine.
12. Sexually transmitted disease such as gonorrhea, chlamydial infections, and genital herpes infections can be transmitted to the fetus at birth. Other sexually transmitted diseases such as syphilis and HIV disease can cross the placenta and affect the developing fetus.

Summary Outline

25.1 Anatomy and Physiology
- A. The genitourinary system is an important portal of entry for pathogens.
- B. The washing action of urination is an important defense mechanism against infections.
- C. Urinary tract infections are more common in women because of the short urethra and the closeness of genital and intestinal tract openings.
- D. The fallopian or uterine tube provides a passageway for microorganisms to enter the abdominal cavity.

25.2 **Normal flora**
- A. **Urethra**: The distal urethra is inhabited by **various microorganisms** including potential pathogens.
- B. **Vagina**: The normal vaginal flora includes **lactobacilli**, which make the vagina more resistant to colonization by pathogens.

25.3 **Urinary system infections**
- A. **Risk factor**: Any condition that impairs normal bladder emptying.
- B. Usually the urinary system becomes infected by **organisms ascending from the urethra**, but it can also be **infected from the bloodstream**.
- C. **Bacterial cystitis** is most often caused by *Escherichia coli* or other enterobacteria from the person's own normal intestinal flora. **Nosocomial urinary infections** are also caused by *Pseudomonas aeruginosa, Serratia marcescens,* and *Enterococcus faecalis*.
- D. **Pyelonephritis** may complicate untreated bladder infection when pathogens ascend through the ureters and involve the kidneys.
- E. **Leptospirosis** is caused by *Leptospira interrogans*, which enters from the bloodstream. Symptoms include **fever, bloodshot eyes** and **pain**.

25.4   **Genital system diseases not transmitted by sexual intercourse**
   A.   **Bacterial vaginosis is the most common cause of vaginal symptoms** including a **gray-white discharge** from the vagina and a **pungent fishy odor**; there is **no inflammation**. The **causative agent** is **unknown**.
   B.   **Vulvovaginal candidiasis** is caused by a yeast, *Candida albicans*, and is the second most common cause of vaginal disorders. Symptoms include **itching, burning, a vulvar rash** and a **thick white discharge. Antibacterial treatment, uncontrolled diabetes** and **oral contraceptives** are predisposing factors.
   C.   **Staphylococcal toxic shock** became widely known with a 1980 epidemic in menstruating women who used a certain kind of **vaginal tampon**. Symptoms include **sudden fever, headache, muscle aches, bloodshot eyes, vomiting, diarrhea**, a **sunburn-like rash** which later peels, and **confusion**. The **blood pressure drops** and, **without treatment, kidney failure** and **death occur**.

25.5   **Sexually transmitted diseases**
   A.   **Incidence in the United States: 15 million new sexually transmitted infections occur each year, including 3 million in teenagers.**
   B.   **Control**: Simple measures for controlling STDs include **abstinence from sexual intercourse**, a **monogamous relationship** with an uninfected person, and consistent **use of latex or polyurethane condoms**.

25.6   **Bacterial STDs**
   A.   **Gonorrhea**, caused by *Neisseria gonorrhoeae*, has been generally **declining in incidence**, but it is still **one of the most commonly reported bacterial diseases**. In **men** the symptoms include **painful urination** and **thick pus draining** from the urethra. In **women** the symptoms tend to be **milder** and are **often overlooked**. Expression of **different surface antigens** allows attachment to different types of cells, **but frustrates development of a vaccine. Inflammatory reaction** to the infection causes scarring that can partially **obstruct the urethra** or cause **sterility** in men and women.
   B.   **Chlamydial STD infections**, caused by *Chlamydia trachomatis*, are **reported more often than any other bacterial disease. Symptoms** and complications of chlamydial infections are **similar to those of gonorrhea, but milder; asymptomatic infections are common** and **readily transmitted**.
   C.   **Syphilis** is caused by *Treponema pallidum*. **Primary syphilis** is noted by a painless firm ulceration called a **hard chancre**; the **organisms multiply and spread throughout the body**. In **secondary syphilis, skin** and **mucous membranes** show **lesions**, which contain the organisms; a **latent period** of months or years separates it from tertiary syphilis. **Tertiary syphilis** is not contagious and is manifest mainly by **damage to the eyes, cardiovascular and central nervous systems**. An inflammatory **necrotizing mass** called a **gumma** can involve any part of the body. Syphilis in **pregnant women** can spread across the placenta to involve the fetus (**congenital syphilis**).
   D.   **Chancroid** is a **widespread** bacterial STD, but it is **not commonly reported** because of difficulties in diagnosis. Caused by *Haemophilus ducreyi*, chancroid is characterized by a **single or multiple soft, tender genital ulcers and enlarged, painful groin lymph nodes**.

25.7   **Viral STDs** are at least as common as bacterial STDs but they are **not** yet **curable**.
   A.   **Genital herpes simplex** is a **very common disease**, important because of the **discomfort** and **emotional trauma** it causes, its **potential for causing death in newborn infants**, its **association with cancer of the cervix** and the **increased risk it poses for HIV infection and AIDS**. Symptoms may include a **group of vesicles with itching, burning or painful sensations**, which break leaving a **superficial ulcer. Local lymph nodes enlarge. Many have few or no symptoms**; some have painful recurrences. The virus establishes a **latent infection in sensory nerves**; it can be transmitted in the absence of symptoms, but the risk is greatest when lesions are present.

B. **Papillomavirus STDs** are **small DNA viruses** that **have not been cultivated in the laboratory**. They are probably **more prevalent than any other kind of STD**. They are **manifest as warts on or near the genitalia and as precancerous lesions**. The **latter are asymptomatic** and can only be detected by medical examination. They are associated with **cervical cancer**.

C. **AIDS is the end stage of disease caused by human immuno-deficiency virus (HIV).** HIV disease is usually first manifest as a **flu-like illness** that develops about six weeks after contracting the virus. An **asymptomatic interval follows** that typically lasts almost ten years during which the **immune system is slowly and progressively destroyed.** **Unusual cancers and infectious diseases** then herald the onset of AIDS. **No vaccine or medical cure is yet available**, but **spread of infection could be significantly slowed by consistent use of condoms and employment of sterile needles by injected-drug abusers**. A marked **reduction in mother to newborn transmission can be achieved with medication**.

25.8 **Protozoan STDs**: Intestinal protozoan diseases such as **giardiasis and cryptosporidiosis** are transmitted by the **fecal-oral route** in those individuals who engage in **oral genital and anal contact** as part of sexual activity.

A. **Trichomoniasis** ("trich") is caused by *Trichomonas vaginalis*. This **often-asymptomatic disease**, may produce symptoms that include itching, burning, swelling and redness of the vagina, with frothy, sometimes smelly, yellow-green discharge and burning on urination. Men have discharge from the penis, burning on urination, sometimes accompanied by painful testes and tender prostate gland.

Terms You Should Know

Congenital syphilis
Ectopic pregnancy
General paresis
Gumma
Hard chancre
Jarisch-Herxheimer reaction

Orchitis
Papanicolaou smear
Papilloma
Pelvic inflammatory disease
  (PID)
Primary syphilis
Secondary syphilis

Sexually transmitted diseases
  (STDs)
Soft chancre
Tertiary syphilis
Vaginosis
Venereal disease

Microorganisms to Know

Normal resident flora

Bacteroides
Corynebacterium
Haemophilus
Lactobacillus sp.
Staphylococcus (coagulase-
  negative)
Streptococcus

Pathogens

Candida albicans
Chlamydia trachomatis
Clostridium perfringens
Enterococcus faecalis
Escherichia coli
Gardnerella vaginalis
Haemophilus ducreyi
Herpes simples virus (HSV)
Human immunodeficiency virus
  (HIV)

Human papillomavirus (HPV)
Leptospira interrogans
Neisseria gonorrhoeae
Pseudomonas aerguinosa
Serratia marcescens
Staphylococcus aureus
Staphylococcus saprophyticus
Treponema pallidum
Trichomonas vaginalis

Diseases to Know

AIDS
Bacterial cystitis
Bacterial vaginosis
Chancroid
Chlamydial infections
Disseminated gonococcal
    infection
Genital herpes simplex

Genital Warts
Gonorrhea
Leptospirosis
Lymphogranuloma venereum
Ophthalmia neonatorum
Pelvic inflammatory disease
    (PID)
Puerperal fever

Pyelonephritis
Syphilis
Toxic shock syndrome
Trachoma
Trichomoniasis
Vulvovaginal candidiasis

Learning Activities

1. Describe the defense mechanisms that help to protect the urinary tract from infection.

2. Explain why the urinary tract is a common site of nosocomial infections.

3.   Give the most common causative agent for:

Disease	Causative agent
Cystitis	
Pyelonephritis	
Leptospirosis	
Bacterial vaginosis	
Vulvovaginal candidiasis	
Toxic shock syndrome	

4.   Describe the following stages of syphilis, list symptoms and state whether the stage is infectious or not.

Stage	Characteristics	Symptoms	Infectious?
Primary			
Secondary			
Tertiary			

5.   Give the characteristics, cause, transmission, treatment, if any, and prevention of the following.

Disease	Characteristics	Cause	Transmission	Treatment and prevention
Chlamydial infections				
Syphilis				
Chancroid				
Genital herpes simplex				
Genital warts				
AIDS				
Trichomoniasis				

6. Match the disease with the distinguishing characteristics.

	Disease		Distinguishing characteristics
	Leptospirosis	a.	Intense, unremitting itching
	Bacterial vaginosis	b.	Painful urination and discharge or asymptomatic
	Vulvovaginal candidiasis	c.	Painful, tiny blisters with underlying redness
	Staphylococcal toxic shock syndrome	d.	Disease has primary, secondary and tertiary stages
	Gonorrhea	e.	Severe damage to the liver and kidneys
	Chancroid	f.	Pungent "fishy" odor
	Genital chlamydial infections	g.	Fever, vomiting, diarrhea, muscle aches, low blood pressure, and a rash that peels
	Syphilis	h.	Generally appears 7 to 14 days after exposure
	Genital herpes simplex	i.	No symptoms or "flu"-like symptoms early in the illness
	Trichomoniasis	j.	Painful, gradually enlarging, soft chancres on or near the genitalia
	Papillomavirus STDs	k.	Women: Itching, burning, swelling, vaginal redness, frothy, sometimes malodorous, yellowish-green discharge Men: Discharge from penis, burning on urination, painful testes, tender prostate.
	AIDS	l.	Warts on the genitalia

7. Describe the prevalence, cause, treatment, pathology and complications of gonorrhea.

Prevalence	
Cause	
Treatment	
Pathology	
Complications	

Self Test

1. Which of the following has resident normal flora?

   a. lower urethra
   b. upper ureter
   c. kidney
   d. upper urethra
   e. lower ureter

2. The urinary tract is protected from infection by

   a. outflow of urine.
   b. antimicrobial organic acids in the urine.
   c. antibodies in the urine.
   d. All of the above are correct.
   e. Only a and b are correct.

3. The most common cause of bladder infections is

   a. *Escherichia coli.*
   b. *Klebsiella pneumoniae.*
   c. *Proteus vulgaris.*
   d. *Serratia marcescens.*
   e. *Pseudomonas aeruginosa.*

4. Which of the following are risk factors for urinary tract infections in women?

   a. A relatively short urethra
   b. The proximity of the urethra opening to that of the gastrointestinal tract
   c. Sexual intercourse
   d. Use of a diaphragm contraception
   e. All of the above are correct.

5. Which of the following is the most common cause of vaginosis?

   a. *Leptospira*
   b. *Escherichia coli*
   c. *Trichomonas vaginalis*
   d. *Chlamydia trachomatis*
   e. *Candida albicans*

6. Which of the following is (are) true for gonorrhea?

   a. It is cause by a Gram-negative diplococcus.
   b. It is the most common reportable sexually transmitted disease.
   c. It causes pelvic inflammatory disease in women.
   d. It is often treated with penicillin or tetracyclines.
   e. All of the above are true.

7. The symptoms of secondary syphilis are

   a. due to the presence of large numbers of bacteria.
   b. due to endotoxins.
   c. due to exotoxins.
   d. associated with concurrent gonococcal infections.
   e. due to immune complexes.

8. Which of the following is true for herpes simplex virus?

   a. Genital herpes is caused by herpes simplex virus type 1 only.
   b. Genital herpes is caused by herpes simplex virus type 2 only.
   c. Genital herpes is caused by either herpes simplex virus type 1 or type 2, but type 1 is more severe.
   d. Genital herpes is caused by either herpes simplex virus type 1 or type 2, but type 2 is more severe.
   e. Genital herpes is caused by either herpes simplex virus type 1 or type 2, which are equal in severity.

9. Which of the following is (are) true for human papillomavirus infections?

   a. These infections are usually spread by sexual intercourse.
   b. A single exposure results in infection 60% of the time.
   c. Asymptomatic individuals can transfer the virus to others.
   d. The virus is associated with cervical cancer.
   e. All of the above are true.

10. Erythromycin is usually used instead of silver nitrate in the prevention of opthalmia neonatorum in newborns because erythromycin can also be used to prevent eye infections due to

    a. *Chlamydia trachomatis.*
    b. *Candida albicans.*
    c. *Trichomonas vaginalis.*
    d. *Treponema pallidum.*
    e. Herpes simplex.

Thought Questions

1. Why are hospitalized patients more likely to get urinary tract infections than other people?

2. Sexually transmitted diseases (STDs) are a major problem. What could public health officials do to decrease the incidence of STDs over a period of time?

Answers to Self Test Questions

1-a, 2-d, 3-a, 4-e, 5-e, 6-e, 7-e, 8-d, 9-e, 10-a

Chapter 26   Nervous System Infections

Overview

While infections of the nervous system are not common, they are very serious. The nervous system does not have normal flora. Infectious agents must enter via the blood, peripheral nerves or wounds or infections of the sinuses, nasal cavity, ears or the mastoid sinus. Treatment of nervous system infections is more difficult than for those of other systems because of the blood-brain barrier, which prevents most antimicrobial medications from entering the nervous system in effective amounts. This chapter presents the most significant nervous system infections.

Learning Objectives

After studying the material in this chapter, you should be able to:

1. Describe how microorganisms gain entrance to the central nervous system.
2. Explain the importance of the blood-brain barrier in relation to treatment of infections of the central nervous system.
3. Describe the characteristics, symptoms, causes, pathogenesis, modes of transmission, prevention and treatment of bacterial meningitis.
4. Describe the characteristics, symptoms, causes, pathogenesis, modes of transmission, prevention and treatment of the following diseases:
   • Listeriosis
   • Leprosy
   • Botulism
5. Describe the characteristics, symptoms, causes, pathogenesis, epidemiology, prevention and treatment of following viral diseases of the nervous system:
   • Meningitis
   • Encephalitis
   • Poliomyelitis
   • Rabies
6. Describe the characteristics, symptoms, causes, pathogenesis, modes of transmission, prevention and treatment of following diseases:
   • Cryptococcal meningoencephalitis
   • African sleeping sickness
7. Define transmissible spongifrom encephalopathy and describe the etiological agent.

Key Concepts

1. The nervous system has no normal flora.
2. Infectious diseases of the nervous system are uncommon, but serious.
3. Infections reach the central nervous system via the blood, the nerves or by spread from local infections.

4. Antimicrobial therapy of nervous system infections is more difficult because many antimicrobial agents are unable to cross the blood-brain barrier.
5. Bacterial meningitis is uncommon, but often life-threatening; bacteria from the respiratory tract are responsible for most infections.
6. Viral meningitis is usually benign, but viral encephalitis often causes permanent damage.
7. Rabies is a widespread a zoonotic disease, which causes a usually fatal encephalitis in humans.
8. Protozoan diseases of the nervous system are rare in the United States, but are more common in tropical areas of the world.
9. Creutzfeld-Jakob disease is a transmissible spongiform encephalopathy that occurs in humans.

Summary Outline

26.1 Infectious agents can reach the **central nervous system** (CNS) via the **bloodstream**, via the **cytoplasm of nerve cell axons** and **by direct extension through bone**.

26.2 **Bacterial diseases of the nervous system**

A. Bacteria can infect the **brain, spinal cord** and **peripheral nerves**, but more commonly they infect the **meninges** and **cerebrospinal fluid**, causing **meningitis**.

B. **Bacterial meningitis is uncommon** and **most victims are children**. In most, but not all victims, the causative bacterium is one commonly found among the **normal upper respiratory flora** of healthy people. *Haemophilus influenzae* was once the leading cause of childhood bacterial meningitis, but it is now mostly controlled by a **vaccine**.

C. **Meningococcal meningitis** is caused by *Neisseria meningitidis*, can result in **shock** and **death** and can occur in **both childhood and adult epidemics. Symptoms are cold-like** and are followed by **abrupt onset of fever, severe headache, pain and stiffness of the neck and back, nausea, and vomiting. Small hemorrhages** into the skin, **deafness**, and **coma** can occur. **Shock** results from the **release of endotoxin** into the bloodstream.

D. **Listeriosis** is caused by *Listeria monocytogenes*, a non spore-forming Gram-positive rod. It is a **food-borne illness** often manifest as **meningitis** in newborn infants and others. The bacterium is **widespread**, commonly contaminates foods such as **unpasteurized milk, cold cuts** and **soft cheeses**, and **can grow in the refrigerator**.

E. **Leprosy (Hansen's disease)** is characterized by invasion of peripheral nerves by the **acid-fast bacillus**, *Mycobacterium leprae*, which **has not been cultivated in vitro**. The disease occurs in **two main forms** depending on the immune status of the patient (1) **Tuberculoid** and (2) **Lepromatous**.

F. **Botulism** is an often fatal type of **food poisoning** that causes severe **generalized paralysis**. The causative bacterium, *Clostridium botulinum*, is an anaerobic Gram-positive rod that **forms heat-resistant spores** that survive canning or other heat-treatment of foods. The spores germinate and the bacteria multiply and release a powerful toxin into the food. **Wound botulism**, caused when *C botulinum*, **colonizes dirty wounds** containing dead tissue, is **rare**.

26.3 **Viral diseases of the nervous system**

A. Most viral nervous system infections are caused by **human enteroviruses** or by the **viruses of certain zoonoses**, but other viruses can **cause infectious mononucleosis, mumps, measles, chickenpox** and **herpes simplex** (cold sores, genital herpes).

B. **Viral meningitis** is much **more common** than bacterial meningitis. It is **generally a mild disease** for which there is **no specific treatment**.

C. **Viral encephalitis** has a **high fatality rate** and often leaves survivors with permanent disabilities. It can be **sporadic or epidemic. Herpes simplex virus** is the most important cause of sporadic encephalitis. **Epidemic encephalitis** is usually caused by **arboviruses**.

LaCrosse encephalitis virus, maintained in *Aedes* mosquitoes and squirrels and chipmunks, is usually the **most frequently reported**.

D. **Poliomyelitis** causes **destruction of motor nerve** cells of the brain and spinal cord leading to **paralysis, muscle wasting** and **failure of normal bone development**.

E. **Rabies** is a **widespread zoonosis** transmitted to humans mainly through the **bite of an infected animal**. Once symptoms appear in an infected human being the disease is **almost uniformly fatal**. Because of the **long incubation period, prompt immunization** with **inactivated vaccine** begun after a rabid animal bite is effective in preventing the disease. **Passive immunization** given at the same time increases the protection.

26.4 Fungal disease of the nervous system

A. **Fungi rarely invade the nervous system of healthy people**, but they can be a **threat to the life of individuals with underlying diseases** such as **diabetes, cancer** and **immunodeficiency**. Treatment of these infections is usually very difficult.

B. **Cryptococcal meningoencephalitis** originates in the lung after a person inhales dust laden with *Filobasidiella (Cryptococcus) neoformans*, **encapsulated yeast**, which **resists phagocytosis** because of its large capsule. The organism is **associated** with soil contaminated with **pigeon droppings**.

26.5 Protozoan diseases of the nervous system

A. **Only a few** free-living protozoa infect the human nervous system.

B. **African sleeping sickness** is caused by *Trypanosoma brucei* and transmitted by the **tsetse fly**. It is a major health problem in **equatorial Africa**. The late stages of the disease are marked by **indifference, sleepiness, coma** and **death**.

26.6 Transmissible spongiform encephalopathies

A. There are a group of rare diseases that cause a sponge-like appearance of the brain tissue due to the loss of neurons.

B. An animal example is mad cow disease.

C. Creutzfeld-Jakob disease is a human example.

D. These diseases are caused by prions that are naked infectious pieces of protein.

Terms You Should Know

Arboviruses
Blood-brain barrier
Cerebrospinal fluid
Encephalitis
Enteroviruses
Infarcts
Meninges

Meningitis
Negri bodies
Neurotoxin
Neurotransmitter
Parasitemia
Petechiae
Picornaviruses

Post-polio syndrome
Prions
Rhabdovirus
Schwartzman phenomenon
Viremia

Microorganisms to Know

Clostridium botulinum
Filobasidiella neoformans
    (Cryptococcus
    neoformans)
Glossina

Haemophilus influenzae
Listeria monocytogenes
Mycobacterium leprae
Naegleria fowleri
Neisseria meningitidis

Poliovirus
Streptococcus pneumoniae
Trypanosoma brucei

Diseases to Know

African trypanosomiasis
Botulism
Creutzfeldt-Jakob disease
Cryptococcal
    meningoencephalitis
Hansen's disease (leprosy)

Lepromatous leprosy
Listeriosis
Meningococcal meningitis
Poliomyelitis
Rabies
Scrapie

Transmissible spongiform
    encephalopathies
Tuberculoid leprosy
Viral meningitis
Primary amebic
    meingoencephalitis

Learning Activities

1. Explain the clinical significance of the blood-brain barrier.

2. What is the major route for microorganisms to get into the central nervous system?

3. List the signs, prevalence and treatment and/or prevention for the following major types of meningitis.

Type of meningitis	Signs	Prevalence	Treatment
Meningococcal meningitis			
Viral meningitis			
Cryptococcal meningoencephalitis			
Streptococcus pneumoniae meningitis			
Haemophilus influenzae meningitis			

4. Describe the symptoms, cause, transmission, pathology and treatment/prevention of listeriosis.

Symptoms	
Cause	
Transmission	
Pathology	
Treatment/ Prevention	

5. Describe the symptoms, cause, transmission, pathology and treatment/prevention of leprosy.

Symptoms	
Cause	
Transmission	
Pathology	
Treatment/ Prevention	

6. Describe the symptoms, cause, transmission, pathology and treatment/prevention of botulism.

Symptoms	
Cause	
Transmission	
Pathology	
Treatment/ Prevention	

7.   Describe the symptoms, cause, transmission, pathology and treatment/prevention of viral meningitis.

Symptoms	
Cause	
Transmission	
Pathology	
Treatment/ Prevention	

8.   Describe the symptoms, cause, transmission, pathology and treatment/prevention of viral encephalitis.

Symptoms	
Cause	
Transmission	
Pathology	
Treatment/ Prevention	

9.   Describe the symptoms, cause, transmission, pathology and treatment/prevention of poliomyelitis.

Symptoms	
Cause	
Transmission	
Pathology	
Treatment/ Prevention	

10. Describe the symptoms, cause, transmission, pathology and treatment/prevention of rabies.

Symptoms	
Cause	
Transmission	
Pathology	
Treatment/ Prevention	

11. Describe the incidence and prevention of arthropod-borne encephalitis.

Incidence	
Prevention	

12. Describe briefly the cause and transmission of the following:

	Cause	Transmission
Cryptococcal Meningoencephalitis		
Primary amebic meningoencephalitis		
African sleeping sickness		

13. Match the means of transmission with the appropriate disease. Answers may be used once, more than once, or not at all.

Answer	Disease	Means of transmission
	Poliomyelitis	a. inhalation of infectious droplets
	Arbovirus encephalitis	b. tsetse fly bite
	Pneumococcal meningitis	c. infection in birth canal
	Leprosy	d. inhalation of dust particles
	Listeria meningitis	e. ingestion of dairy products
	Cryptococcal meningitis	f. animal bite
	African sleeping sickness	g. direct contact
	Meningococcal meningitis	h. fecal-oral route
	Rabies	i. mosquito bite
	Botulism	j. ingestion of contaminated canned food

Self Test

1. Which of the following is true about the normal flora of the nervous system?

   a. Only transient organisms are present.
   b. Microorganisms are present only in portions of the central nervous system
   c. The nervous system does not have normal flora.
   d. Microorganisms are present only in portions of the peripheral nervous system.
   e. Both a and c are correct.

2. Infections of the central nervous system are difficult to treat because

   a. many of the antimicrobial drugs cannot pass through the blood-brain barrier.
   b. nerve tissue does not regenerate so healing is very difficult.
   c. most infections are caused by fungi and they are difficult to treat.
   d. many of the antimicrobial drugs also affect the brain and cause damage.
   e. most infections occur only after extensive brain damage.

3. Which of the following bacteria is the principal cause of meningitis in adults?

   a. *Streptococcus pneumoniae*
   b. *Haemophilus influenzae*
   c. *Neisseria meningitidis*
   d. *Listeria monocytogenes*
   e. *Streptococcus pyogenes*

4. Which of the following types of meningitis is often transmitted by food?

   a. *Streptococcus pneumoniae* meningitis
   b. *Haemophilus influenzae* meningitis
   c. *Neisseria meningitidis* meningitis
   d. *Listeria monocytogenes* meningitis
   e. None of these are transmitted by the food.

5. In rabies infections

   a. treatment after infection is possible now because of new viral vaccines.
   b. most cases in the United States today are in wild animals.
   c. the incubation period is often long and dependent on the site of the bite and the dose of virus.
   d. Both a and b are correct.
   e. Both b and c are correct.

6. Leprosy or Hansen's disease is best characterized as

   a. a highly contagious disease.
   b. an infection of the peripheral nervous system.
   c. an acute infection.
   d. an infection of the central nervous system.
   e. Both a and c are correct.

7. *Streptococcus pneumoniae, Haemophilus influenzae,* and *Neisseria meningitidis* seldom cause meningitis in newborns because

   a. most mothers have antibodies against them and these are transferred across the placenta.
   b. newborns lack the appropriate adhesions for the infections.
   c. most mothers have antigens against them and these are transferred across the placenta.
   d. these organisms lack the appropriate adhesions for the infections.
   e. These organisms do often cause meningitis in newborns.

8. Which of the following microorganisms causes meningoencephalitis?

   a. *Streptococcus pneumoniae*
   b. *Filobasidiella (Cryptococcus) neoformans*
   c. *Haemophilus influenzae*
   d. *Naegleria fowleri*
   e. Both b and c are correct.

9. The causative agent of spongiform encephalopathies appears to be

   a. a virus.
   b. a prion.
   c. a viroid.
   d. a protozoan.
   e. a fungus.

10. The tsetse fly is the vector for which of the following diseases?

    a. meningococcal meningitis
    b. poliomyelitis
    c. arbovirus encephalitis
    d. African sleeping sickness
    e. None of the above.

Thought Questions

1. Explain why meningitis and encephalitis are difficult to treat. How do we get around these problems?

2. Describe how a physician might deal with a person who was bitten by a wild raccoon. What diseases and treatments should be considered?

Answers to Self Test Questions

1-c, 2-a, 3-a, 4-d, 5-e, 6-b, 7-a, 8-b, 9-b, 10-d

Chapter 27   Wound Infections

Overview

Wounds are breaks in the skin that allow microorganisms, either overt pathogens or opportunists, to enter the body and cause infection. Any wound may result in significant disease. Some of these diseases result from the spread of invasive microorganisms throughout the body. Others result from the spread of exotoxins that can cause tissue destruction. This chapter presents the presents the major aspects of wound infections.

Learning Objectives

After studying the material in this chapter, you should be able to:

1. Define wound infections and describe their possible consequences.
2. List the types of wounds and describe abscesses and anaerobic wounds.
3. Identify the most common bacterial causes of wound infections.
4. Explain why *Staphylococcus aureus* is an important cause of wound infections in humans.
5. List and describe the cause, pathology, transmission, prevention and treatment of the following kinds of infections:
   - Staphylococcal wound infections
   - Streptococcal wound infections
   - Necrotizing fasciitis
   - *Pseudomonas* wound infections
   - Tetanus
   - Gas gangrene
   - Actinomycosis
   - Bite wounds
   - Cat scratch disease
   - Rat bite fever
6. Explain why the causative agent of tetanus causes significant disease even though it is non-invasive.
7. Explain how *Clostridium perfringens* causes disease.
8. List the major fungal wound infections and give the cause, pathology, prevention and treatment of each.
9. Explain why human bites are so dangerous.

Key Concepts

1. Wounds expose tissues that can be infected by some microorganisms.
2. Anaerobic conditions can be created in wounds by the presence of dead tissue and foreign material, resulting in colonization by microorganisms that would not normally grow in those locations.
3. Bacteria that have little capability for invasion can produce serious disease through the production of exotoxins that can be absorbed into the blood and transported to other parts of the body.

4. Since many different kinds of microorganisms may infect wounds, diagnostic and treatment methods vary greatly.
5. *Staphylococcus aureus* is considered the most significant cause of wound infections because it is commonly carried by humans, can be easily transmitted from person to person, and possesses a number of virulence factors.
6. *Streptococcus pyogenes* strains are uncommon in wound infections, but cause significant disease when they do occur.
7. *Pseudomonas aeruginosa* is a major cause of nosocomial infections.
8. Bacterial exotoxins from different species of bacteria may have the same mode of action, but cause distinctly different diseases because the toxins attach to and enter different types of cells.
9. Tetanus is caused by anaerobic noninvasive spore-forming bacteria, which produce a neurotoxin that causes significant tissue damage.
10. Gas gangrene is caused by anaerobic spore-forming bacteria that live on dead tissues. The bacteria cause disease by spreading and releasing toxins.
11. Animal bites and scratches can transmit bacteria such as *Pasteurella multocida* and *Bartonella henselae.*
12. *Candida albicans*, often part of the normal flora, can infect burns and wounds.
13. *Sporothrix schenckii* may cause fungal wound infections in persons whose occupations expose them to splinters and sharp vegetation.

Summary Outline

27.1 **Wounds** expose components of tissues to which pathogens specifically attach.
   A.   **Healing**: Wounds heal by forming **granulation tissue** that normally fills the defect and then contracts to minimize scar tissue.
   B.   **Thermal burns** often present large areas of dead tissue devoid of competing organisms and body defenses providing ideal conditions for microbial growth.
   C.   Wound **abscess formation localizes an infection** within tissue to prevent its spread. An abscess involves a collection of **pus**, which is **composed of leukocytes, components of tissue breakdown and infecting organisms**.
   D.   **Anaerobic conditions** are likely to occur in **wounds containing dead tissue** or **foreign material**, and **those with a narrow opening to the air**. These conditions permit infection by particularly dangerous pathogens.
27.2 **Common bacterial wound infections**
   A.   Possible **consequences of wound infections** include (1) **delayed healing**, (2) **abscess formation** and (3) **extension of infection or toxins** into adjacent tissues or the bloodstream. Infections can cause **surgical wounds to split open**, and they can spread to **create biofilms** on artificial devices.
   B.   **Staphylococcal wound infections**: Staphylococci, usually *Staphylococcus aureus* or *S. epidermidis*, are the leading cause of wound infections. *Staphylococcus aureus* possesses many **virulence factors**; occasional strains release a toxin that causes **toxic shock syndrome**. *Staphylococcus epidermidis* is **less virulent** but **can form biofilms** on blood vessel catheters and other devices.
   C.   *Streptococcus pyogenes* (Group A, β-hemolytic streptococcus) (flesh-eaters) causes "**strep throat**," **scarlet fever**, **wound infections** and other conditions. **Necrotizing fasciitis-causing strains produce exotoxin B**, a protease thought to be responsible for tissue destruction.
   D.   *Pseudomonas aeruginosa,* an **aerobic Gram-negative rod** with a single polar flagellum, is an **opportunistic pathogen** widespread in the environment and a cause of both **nosocomial infections** and those acquired outside the hospital.

27.3 Diseases due to anaerobic wound infections

A. **Tetanus (Lockjaw) is an often fatal disease** characterized by **sustained, painful, cramp-like spasms** of one or more muscles. It is caused by an **exotoxin, tetanospasmin**, produced by *Clostridium tetani*, a **noninvasive, anaerobic Gram-positive rod**. This toxin renders the nerve cells that normally inhibit muscle contraction inactive by **blocking release of their neurotransmitter**. Tetanus can be prevented by **active immunization with toxoid** (inactivated tetanospasmin).

B. **Gas gangrene (clostridial myonecrosis)** is usually caused by the anaerobe *Clostridium perfringens*. **Symptoms** begin abruptly with **pain, swelling**, a **thin brown bubbly discharge**, and **dark blue mottling** of the tightly stretched overlying **skin**. The **toxin causes tissue necrosis**; hydrogen and carbon dioxide gases are produced from fermentation of amino acids and glycogen in the dead tissue. Since there is **no vaccine or toxoid, prevention depends on prompt medical care of dirty wounds. Treatment depends on urgent surgical removal of dead and infected tissue** and may require amputation.

C. **Actinomycosis (lumpy jaw)** is a **chronic, slowly progressive disease** characterized by **repeated swellings, discharge of pus** and **scarring**, usually of the face and neck. The causative agent is *Actinomyces israelii*, a member of the **normal mouth, intestinal and vaginal flora** that enters tissues with wounds such as those with dental and intestinal surgery. The organism is **slow growing**; treatment must be continued for weeks or months.

27.4 **Bacterial bite wound infections**

A. *Pasteurella multocida*, a small Gram-negative rod, can infect bite wounds inflicted by a **number of animal species**. Specific opsonizing antibody permits killing of the bacteria by phagocytes.

B. **Cat scratch disease** is the most common cause of chronic localized lymph node enlargement in children. Caused by *Bartonella henselae*, it begins with a pimple at the site of bite or scratch, followed by **enlargement of local lymph nodes,** which often become pus-filled. **Most** individuals with cat scratch disease **recover without treatment**.

C. **Streptobacillary rat bite fever** is characterized by **relapsing fevers, head and muscle aches,** and **vomiting**, following a rat bite. A **rash and joint pain** often develop. It is usually caused by *Streptobacillus moniliformis*, a highly **pleiomorphic Gram-negative rod** that produces cell wall-deficient variants called **L-forms**.

D. **Human bite wound infections** are usually caused by members of the **normal flora** acting synergistically, including **anaerobic streptococci, fusiforms, spirochetes** and *Bacteroides* **sp.** often with *Staphylococcus aureus*.

27.5 **Fungal wound infections** are unusual in economically developed countries except for *Candida albicans* infections of burns and other wounds in individuals receiving antibacterial therapy.

A. **Madura foot** occurs in many impoverished areas of the world where people do not wear shoes.

B. **Sporotrichosis (rose gardeners' disease)** is a **chronic fungal disease** caused by the dimorphic fungus *Sporothrix schenckii*. The fungus is distributed worldwide in **tropical and temperate climates** and is usually introduced into wounds caused by **thorns or splinters**. Symptoms include **painless ulcerating nodules that develop along the course of a lymphatic vessel**.

Terms You Should Know

α-toxin	L-forms	Synergistic infection
Abscesses	MRSAs	Tetanospasmin
Granulation tissue	Pyogenic	Tetanus immune globulin (TIG)

Microorganisms to Know

Actinomyces israelii
Bartonella henselae
Candida albicans
Clostridium perfringens

Clostridium tetani
Pasteurella multocida
Pseudomonas aeruginosa
Sporothrix schenckii

Staphylococcus aureus
Staphylococcus epidermidis
Streptobacillus moniliformis
Streptococcus pyogenes

Diseases to Know

Actinomycosis
Cat scratch disease
Gas gangrene (clostridial
    myonecrosis)
Haverhill fever
Human bite infection

Madura foot
Necrotizing fasciitis
Pasteurella multocida bite
    wound infection
Pseudomonas aeruginosa
    infection

Sporotrichosis
Staphylococcal wound infection
Streptobacillary rat bite fever
Tetanus

Learning Activities

1. List four factors that determine whether microorganisms in a wound will cause disease.

1.	
2.	
3.	
4.	

2. List and describe the four types of wounds.

Type of wound	Description

3. List three possible consequences of wound infections.

1.	
2.	
3.	

4. List the causative agent of the following diseases.

Disease	Causative agent
Staphylococcal wound infections	
Streptococcal wound infections	
Necrotizing fasciitis	
Pseudomonas wound infections	
Tetanus	
Gas gangrene	
Actinomycosis	
Animal bites	
Cat scratch disease	
Rat bite fever	
Yeast infections	
Madura foot	
Sporotrichosis	

5. Describe how the following infections caused by the following microorganisms may be prevented and treated.

Infection	Prevention	Treatment
Staphylococcus aureus		
Streptococcus Group A		
Pseudomonas aeruginosa		
Clostridium tetani		
Clostridium perfringens		
Actinomycosis israelii		
Pasteurella multocida		
Bartonella henselae		
Candida albicans		
Sporothrix schenckii		

6. List the mode of transmission for the following diseases.

Disease	Mode of transmission
Staphylococcal wound infections	
Streptococcal wound infections	
Necrotizing fasciitis	
Pseudomonas wound infections	
Tetanus	
Gas gangrene	
Actinomycosis	
Animal bites	
Cat scratch disease	
Rat bite fever	
Candida albicans infections	
Madura foot	
Sporotrichosis	

7. Describe the pathogenesis of wound infections caused by the following microorganisms.

Microorganism	Pathogenesis
Staphylococcus aureus	
Clostridium tetani	
Clostridium perfringens	

8. Explain how *Clostridium tetani* causes tetanus even though it is non-invasive.

9. Explain how *Clostridium perfringens* causes disease.

10. Name the microorganism most often associated with the following.

Type of wound	Causative agent
Trauma wound	
Surgical wound	
Burn	
Animal bite	
Cat scratch	
Human bite	
Thorn puncture wound	

11. Why are human bite wounds significant?

Self Test

1. The major cause of death due to infection in burn patients is

    a. *Staphylococcus aureus.*
    b. *Staphylococcus epidermidis.*
    c. *Pseudomonas aeruginosa.*
    d. *Pasteurella multocida.*
    e. *Candida albicans.*

2. The most important cause of wound infections is

    a. *Staphylococcus aureus.*
    b. *Staphylococcus epidermidis.*
    c. *Pseudomonas aeruginosa.*
    d. *Streptococcus pyogenes.*
    e. *Candida albicans.*

3. Which of the following microorganisms produces a greenish discoloration of a wound?

    a. *Streptococcus pyogenes*
    b. *Staphylococcus epidermidis*
    c. *Staphylococcus aureus*
    d. *Pasteurella multocida*
    e. *Pseudomonas aeruginosa*

4. Which of the following microorganisms causes cat scratch fever?

    a. *Bartonella henselae*
    b. *Staphylococcus epidermidis*
    c. *Staphylococcus aureus*
    d. *Pasteurella multocida*
    e. *Pseudomonas aeruginosa*

5. Which of the following microorganisms is introduced into wounds by thorns and splinters?

   a. *Pseudomonas aeruginosa*
   b. *Staphylococcus aureus*
   c. *Pasteurella multocida*
   d. *Candida albicans*
   e. *Sporothrix schenckii*

6. Which of the following microorganisms is a non-invasive exotoxin producer?

   a. *Clostridium tetani*
   b. *Clostridium perfringens*
   c. *Bartonella henselae*
   d. *Bacteroides sp.*
   e. *Pasteurella multocida*

7. The patient's own body is the most common source for which of these bacteria?

   a. *Streptococcus pyogenes*
   b. *Staphylococcus epidermidis*
   c. *Staphylococcus aureus*
   d. *Pasteurella multocida*
   e. *Pseudomonas aeruginosa*

8. Which of these bacteria spread disease to healthy tissue with an alpha-toxin?

   a. *Clostridium tetani*
   b. *Clostridium perfringens*
   c. *Staphylococcus aureus*
   d. *Pasteurella multocida*
   e. *Streptobacillus moniliformis*

9. Which of the following is the most common cause of myonecrosis?

   a. *Clostridium tetani*
   b. *Clostridium perfringens*
   c. *Staphylococcus aureus*
   d. *Pasteurella multocida*
   e. *Streptobacillus moniliformis*

10. Which of the following organisms most often results in disease from animal bites?

   a. *Sporothrix schenckii*
   b. *Clostridium perfringens*
   c. *Pseudomonas aeruginosa*
   d. *Pasteurella multocida*
   e. *Streptobacillus moniliformis*

Thought Questions

1. Describe the conditions that lead to the development of gas gangrene. Explain why antibiotics must be promptly administered to effectively treat this disease.

2. Explain why anaerobic wounds are significant; describe treatment and the rationale for treatment.

Answers to Self Test Questions

1-c, 2-a, 3-e, 4-a, 5-e, 6-a, 7-c, 8-b, 9-b, 10-d

Chapter 28   Blood and Lymphatic Infections

Overview

Blood and lymphatic infections are potentially serious because they can carry microorganisms throughout the body, resulting in systemic infections. Bacteremia is the presence of bacteria in the blood, which under normal circumstances is sterile. Septicemia infers not only presence of bacteria, but actively dividing bacteria in the blood. Malaria, a protozoan disease of the blood, which has been a problem since ancient times, is still the most common infectious disease in the world killing more than a million people per year. The plague, while now uncommon, was the cause of the worst recorded epidemic up to modern times, killing about one-fourth of the population of Europe in the Middle Ages. This chapter presents the major diseases of the blood and lymphatic systems.

Learning Objectives

After studying the material in this chapter, you should be able to:

Relate the nature of circulatory and lymphatic systems to the occurrence of generalized infectious disease.
Explain why diseases of the circulatory and lymphatic systems are often life- threatening.
Describe the symptoms, causes, pathology, occurrence, prevention and treatment of the following diseases.
- Subacute bacterial endocarditis
- Gram-negative septicemia
- Tularemia
- Brucellosis
- Infectious mononucleosis
- Yellow fever

Describe the history, symptoms, cause, pathology, forms of the disease, modes of transmission, prevention and treatment of plague.
Explain why malignant tumors associated with the Epstein-Barr virus occur in a limited area while the virus is prevalent worldwide.
Explain why malaria is still a major disease with worldwide implications.
Describe the symptoms, cause, pathology, mode of transmission and treatment of malaria.
Describe the life cycle of malaria and relate the stages of the cycle to prevention of the disease.

Key Concepts

1. A systemic infection represents failure of the body's defense mechanism to keep an infection localized.
2. Systemic infections threaten the delivery of oxygen and nutrients to tissues resulting in generalized and often life-threatening disease.
3. The nature of the circulatory system and the lymphatic system facilitates the spread of infectious disease throughout the body.

4. The inflammatory response, although important in localizing infections, can result in a life-threatening situation when generalized.
5. Tularemia is a bacteremic disease of wild animals that can be transmitted to humans by exposure to blood or tissues, or by biting insects.
6. Brucellosis is a disease of domestic animals that can be transmitted to humans by contact with diseased carcasses or by drinking unpasteurized milk.
7. Plague is caused by *Yersinia pestis* and is contracted from the bite of an infected rat flea. The disease is endemic in rodents in the western United States and has the potential to spread rapidly from person to person by the respiratory route. Three forms of the disease exist: (a) bubonic plague characterized by enlarged lymph nodes called buboes, (b) septicemic plague in which the bacteria is found in the blood, and (c) pneumonic plague that occurs when the lungs are infected. The third form is both highly contagious and lethal.
8. Infectious mononucleosis has worldwide incidence and is transmitted from person to person by saliva. The causative agent is the Epstein-Barr virus (EBV), which is also associated with malignancies in limited areas, and with AIDS.
9. Yellow fever is the classic hemorrhagic fever, and is caused by the yellow fever virus, which is an enveloped single-stranded RNA virus of the flavirius family. It is a tropical disease transmitted by mosquitoes. Monkeys are the reservoir. While there is no effective treatment, there is a highly effective live attenuated viral vaccine available.
10. Malaria is a protozoan disease with worldwide implications. It is caused by one of four species of *Plasmodium* transmitted by the *Anopheles* mosquito. Symptoms include recurrent bouts of violent chills and fever alternating with healthy feelings. Treatment with chloroquine and primaquine has been mainly ineffective. Prevention by eradication of the mosquito vectors has been successful in some places, but not in all. Vaccines are under development, but so far have proven ineffective.

Summary Outline

28.1 Anatomy and Physiology
   A. The cardiovascular system is a transportation system composed of the heart, blood vessels, blood, lymph nodes, lymph vessels, lymph and capillaries
   B. Arterosclerotic lesions are often colonized by *Chlamydia pneumonia*, whose signifiance in the disease process is not yet understood.
28.2 **Bacterial diseases of the blood vascular system**
   A. Bacteria circulating in the bloodstream can colonize the inside of the heart, and cause collapse of the circulatory system and death. Infections of the heart valves and lining of the heart are called endocarditis. Illness resulting from circulating pathogens is called septicemia.
   B. **Acute bacterial endocarditis** is caused when virulent bacteria enter the bloodstream from a **focus of infection; normal heart valves are commonly infected and destroyed**.
   C. **Subacute bacterial endocarditis** (SBE) is commonly caused by organisms of little virulence, including **oral streptococci** and *Staphylococcus epidermidis*. Infection usually begins on **structural abnormalities of the heart**.
   D. **Gram-negative septicemia** is commonly a **nosocomial illness**. Often affects individuals have **serious underlying illnesses** such as cancer and diabetes. A **common complication is shock** precipitated by **release of endotoxin** from the bacteria.
28.3 **Bacterial diseases involving the lymph nodes and spleen**
   A. **Tularemia, brucellosis** and **plague** involve the **mononuclear-phagocyte system** and are characterized by **enlargement of the lymph nodes and spleen**.
   B. The causative organisms **grow within phagocytes**, protected from antibody.

C.   **Tularemia (rabbit fever)** is usually transmitted from wild animals to humans by exposure to the animals' blood or by insects and ticks. The cause is the Gram-negative aerobe *Francisella tularensis*.

D.   **Brucellosis (undulant fever)**, caused by *Brucella melitensis*, usually **acquired from cattle or other domestic animals**. The organisms can infect via mucous membranes and minor skin injuries.

E.   **Plague (black death)**, once pandemic, now persists endemically in **rodent populations**, including those in many Western states of the U.S. It is caused by *Yersinia pestis*, an enterobacterium with many **virulence factors, chromosomally or plasmid coded, which interfere with phagocytosis and immunity. Bubonic plague** is **transmitted to humans by fleas** and **pneumonic plague** is transmitted person-to-person. Untreated, **bubonic plague** has a **mortality rate of 50-80%** and **pneumonic plague** has a mortality rate of **almost 100%**

28.4   **Viral disease of the lymphoid or blood vascular systems**

A.   **Infectious mononucleosis** (mono, kissing disease) is caused by **Epstein-Barr virus (EBV)**, which establishes a **lifelong latent infection** of B lymphocytes. Incidence is high in 15- to 24-year olds.

B.   **Yellow fever** is a **zoonosis** of mosquitoes and monkeys that exists mainly in **tropical jungles**. It can become epidemic in humans where a suitable *Aedes* **mosquito vector** is present. The disease involves the **heart and blood vessels** and is characterized by **fever, jaundice** and **hemorrhaging**. There is a highly effective **live attenuated vaccine** available.

28.5   **Protozoan diseases**

A.   **African sleeping sickness** is present over much of tropical Africa.

B.   **Malaria**, caused by **four species of *Plasmodium*** and transmitted from person to person by the bite of *Anopheles* mosquitoes, is the **most widespread of the all serious infectious diseases**, mainly found in impoverished warm regions of the world. The **life cycle is complex**; different forms of the organism invade different body cells and have **different susceptibility to antimalarial medication**.

Terms You Should Know

Aneurysm	Exoerythrocytic cycle	Sporozoite
Bacteremia	Heterophile antibody	Systemic
Buboes	Lymphangitis	Trophozoite
Disseminated intravascular coagulation (DIC)	Merozoite	Viremia
Endocarditis	Osteomyelitis	
	Septic shock	

Diseases to Know

Acute bacterial endocarditis	Burkitt's lymphoma	Septicemia
African sleeping sickness	Infectious mononucleosis	Subacute bacterial endocarditis
Brucellosis (undulant fever)	Malaria	Tularemia
Bubonic plague	Pneumonic plague	Yellow fever

Microorganisms to Know

α-hemolytic viridans streptococci	Epstein-Barr virus (EBV)	Staphylococcus aureus
Aedes aegypti	Escherichia coli	Staphylococcus epidermidis
Anopheles	Francisella tularensis	Streptococcus pneumoniae
Bacteroides sp.	Plasmodium falciparum	Tropheryma whippelii
Brucella melitensis	Plasmodium malariae	Trypanosoma brucei
Chlamydia pneumoniae	Plasmodium ovale	Yellow fever virus
Coxiella brunetii	Plasmodium vivax	Yersinia perstis
	Pseudomonas aeruginosa	

Learning Activities

1. Define and differentiate among:

Bacteremia	
Septicemia	
Septic shock	

List the characteristics of septicemia including symptoms, causes, and manifestations.

Symptoms	
Causes	
Manifestations	

Describe the characteristics of tularemia including symptoms, pathology and mode of transmission.

Symptoms	
Pathology	
Mode of transmission	

Indicate the causative agent of the following diseases:

Disease	Causative agent
Subacute bacterial endocarditis	
Tularemia	
Brucellosis	
Plague	
Infectious mononucleosis	
Yellow fever	
Malaria	

Describe the symptoms and mode of transmission of the following diseases:

Disease	Symptoms	Mode of transmission
Brucellosis		
Bubonic plague		
Pneumonic plague		
Infectious mononucleosis		
Yellow fever		

Describe malaria. Explain how it is diagnosed.

Symptoms	
Pathology	
Mode of transmission	
Causative agent	
Reservoir	
Means of diagnosis	

Self Test

1. *Streptococcus epidermidis* and alpha-hemolytic streptococcus are causative agents for which of the following diseases?

    a.   subacute bacterial endocarditis
    b.   septicemia
    c.   tularemia
    d.   brucellosis
    e.   plague

2. Which of the following diseases is endemic in rodent populations worldwide?

    a.   tularemia
    b.   brucellosis
    c.   yellow fever
    d.   plague
    e.   malaria

3. Enlarged lymph nodes called buboes are a feature of which of the following diseases?

    a.   malaria
    b.   yellow fever
    c.   brucellosis
    d.   plague
    e.   tularemia

4. Fleas are a vector for

    a.   malaria.
    b.   plague.
    c.   tularemia.
    d.   brucellosis.
    e.   yellow fever.

5. Shock is a common feature of

    a.   subacute bacterial endocarditis.
    b.   septicemia.
    c.   tularemia.
    d.   brucellosis.
    e.   plague.

6. Which of the following diseases has a relatively high incidence in 15- to 24-year olds?

    a.   subacute bacterial endocarditis
    b.   infectious mononucleosis
    c.   brucellosis
    d.   tularemia
    e.   None of the above.

7. Recurrent bouts of fever are a sign of which of the following diseases?

    a.   malaria
    b.   septicemia
    c.   tularemia
    d.   yellow fever
    e.   brucellosis

8. Which of the following diseases are zoonoses?

    a.   tularemia
    b.   brucellosis
    c.   plague
    d.   All of the above are correct.
    e.   Only a and c are correct.

9. Symptoms of high fever, headache, jaundice, black vomit, and hemorrhages into the skin are indicative of which of the following diseases?

    a.   brucellosis
    b.   tularemia
    c.   infectious mononucleosis
    d.   malaria
    e.   yellow fever

10. Which of the following diseases has a mosquito as a vector?

    a.   yellow fever
    b.   plague
    c.   malaria
    d.   tularemia
    e.   Both a and c are correct.

Thought Questions

1. Explain why infectious mononucleosis caused by the Epstein-Barr virus has worldwide incidence, but is associated with malignances in only a limited area.

2. Why was the plague such a major problem in the Middle Ages, but is a relatively uncommon problem now?

Answers to Self Test Questions

1-a, 2-d, 3-d, 4-b, 5-b, 6-b, 7-a, 8-d, 9-e, 10-e

Chapter 29   HIV Disease and Complications of Immunodeficiency

Overview

Since acquired immunodeficiency syndrome (AIDS) was first recognized in 1981, this disease has become a major worldwide problem of epidemic proportions. The disease is actually a late manifestation of human immunodeficiency virus (HIV) infection. AIDS is the result of a failure of the immune system to protect the HIV-infected individual against pathogens and opportunistic infections. It is also associated with some malignancies. This chapter presents HIV, its transmission, infection and outcomes.

Learning Objectives

After studying the material in this chapter, you should be able to:

1. Describe the global scope of HIV disease in terms of incidence and prevalence.
2. Identify the causative agent of HIV disease.
3. Differentiate between HIV disease and the end stage immunodeficiency known as AIDS.
4. List the AIDS-defining conditions.
5. Describe the HIV genome and its gene products.
6. Describe the major symptoms of HIV disease.
7. Explain how HIV infects certain types of cells and specifically destroys the immune system, causing disease in humans.
8. Explain how HIV is transmitted.
9. Describe how HIV disease can be prevented.
10. Describe how HIV disease can be treated.
11. List the major risk factors for HIV disease.
12. Explain why vaccines have not been successful in the prevention of HIV disease and describe the prospects of an effective vaccine in the future.
13. List and describe the kinds of tumors associated with HIV disease.
14. List and describe the major infectious complications of acquired immunodeficiency.

Key Concepts

1. The signs and symptoms of AIDS are caused primarily by opportunistic infections and tumors that complicate HIV disease.
2. HIV is transmitted mainly by sexual contact, blood or blood products, and from mother to child around the time of childbirth.
3. HIV infectious various body cells, but most importantly $CD4^+$ T lymphocytes and antibody-presenting cells. T cells are killed and the number slowly declines resulting in an immune system that can no longer resist infection or development of tumors.
4. The worldwide incidence of AIDS could be significantly altered by changes in human behavior.
5. While antiretroviral therapy has been successful in improving the condition of many patients with AIDS, it is unlikely to alter the incidence of the disease.

6. It has been found that some DNA viruses, along with cofactors, are strongly associated with the development of malignant tumors in patients with HIV disease.
7. Individuals with immunodeficiencies including AIDS are much more susceptible to infectious diseases, including reactivation of latent infections, than healthy people.
8. Prevention of infectious disease in AIDS patients is essential to improved management and prognosis of these patients.

Summary Outline

AIDS was first recognized in 1981 in sexually promiscuous homosexual men. It has claimed over **420,000 lives in the United States**, and become the **leading cause of death** in those **25 to 44** years of age. Worldwide, **33 million people** are infected with an AIDS-causing virus.

29.1. **Human immunodeficiency virus (HIV) and AIDS**

    A. **"AIDS defining conditions"** include **unusual tumors** or **serious infections** by agents that normally have **little virulence**. These conditions usually reflect **immunodeficiency**.

    B. **HIV disease: Acquired immunodeficiency syndrome (AIDS)** is **a late manifestation** of human immunodeficiency virus (HIV) disease. **Early symptoms** of the acute retrovirus syndrome are **"flu-like"** and often occur six days to six weeks after infection by HIV; these **symptoms subside without treatment**. HIV disease **progresses** and can be transmitted to others. The **asymptomatic period ends with the appearance of tumors, or the onset of immunodeficiency (AIDS).**

    C. **Causative agent** is the **Human immunodeficiency virus type 1 (HIV-1)**, a **single-stranded RNA virus** of the **retrovirus family**.

    D. **Transmission** is mainly by **sexual intercourse, by blood contaminated hypodermic needles,** and **from mother to fetus or newborn**.

    E. **Infection of cells**: HIV infects **T helper (Th) lymphocytes** and **macrophages**. Inside the host cell a **DNA copy of the viral genome** is made through the action of **reverse transcriptase**, a complementary DNA strand is made, and the **double-stranded DNA is inserted into the host genome as a provirus** by viral integrase (IN).

    F. **Antimicrobial medications** are used to prevent and treat opportunistic infections, and combinations of antiretroviral medications (HAART) are used to treat HIV disease.

    G. **HIV vaccine prospects**: No approved vaccine is currently available, but a number of prospects have moved from animal to human trials.

29.2 **Malignant tumors** that complicate acquired immunodeficiencies

    A. **Kaposi's sarcoma** is a tumor arising from blood or lymphatic vessels. There is markedly **increased incidence among immunodeficient individuals**. Infection by **human herpes virus 8 appears to be required** for the tumor to develop.

    B. **B lymphocytic tumors of the brain** are **malignant tumors (lymphomas)** that arise from lymphoid cells. There is strong evidence exists that **EB virus plays a causative role** in these tumors.

    C. There is an **increased rate of anal, genital and cervical carcinoma** in people with HIV disease; these tumors arise from squamous epithelial cells and are strongly associated with **human papillomaviruses (HPV)**, which are transmitted by sexual intercourse.

29.3 Infectious complications of acquired immunodeficiency

    A. **Infections that occur in healthy individuals** also occur and **produce more severe disease in those with immunodeficiency. Latent infections** such as those by *Mycobacterium tuberculosis* and herpes viruses are **commonly activated**, and **organisms that are rarely capable of causing disease** in healthy individuals **can be life-threatening**.

B.   **Pneumocystosis**, caused by *Pneumocystis carinii*, is **widespread** and **usually asymptomatic**. It was the **most common cause of death among AIDS sufferers** before effective preventive regiments were developed. Symptoms develop slowly, with gradually **increasing shortness of breath** and **rapid breathing**.

C.   **Toxoplasmosis, rare among healthy people**, can be a serious problem for those with cancer, organ transplantation and HIV disease. The disease can also be congenital. **Symptoms**, which usually disappear without treatment, include **sore throat, fever, enlarged lymph nodes and spleen**, and sometimes a **rash**. Infections early in pregnancy can cause **miscarriage** or **birth defects**. **Infections later in pregnancy** are usually milder but can result in **epilepsy, mental retardation** or **recurrent retinitis** in the child. The causative agent is *Toxoplasma gondii*, a tiny protozoan that undergoes sexual reproduction in the intestinal epithelium of cats. **Oocysts** are discharged in cat feces and can infect humans through **contaminated drinking water** or **eating rare meat**.

D.   **Cytomegalovirus**, an **enveloped, double-stranded DNA virus**, is a common cause of **impaired vision in people with AIDS**. The virus can exist in a **latent form**, a **slowly replicating form**, or a **fully replicating form**. **Coinfection with HIV** results in **fully productive infection** and tissue death. **No vaccine** is available. The antiviral medication **ganciclovir** can be given to prevent CMV retinitis.

E.   **Mycobacterial diseases** are commonly caused **by *Mycobacterium tuberculosis*** and *Mycobacterium avium* **complex (MAC)** organisms. **Normal** people usually get **asymptomatic** or **mild infections** with MAC organisms, but **immunodeficient patients** may have **fever, drenching sweats, severe weight loss, diarrhea and abdominal pain**. MAC organisms enter the body via the lungs and gastrointestinal tract, are taken up by **macrophages, but resist destruction**, and are carried to all parts of the body. In people who are immunodeficient, MAC **organisms multiply without restriction**, producing massive numbers of organisms in blood, intestinal epithelium and tissues. There are **no generally effective measures** available, but **prophylactic medication is advised for severely immunodeficient patients**.

Terms You Should Know

Acute retroviral syndrome (ARS)
AIDS-related complex (ADC)
Carcinoma
CD4 antigen
HAART
Hairy leukoplakia
Lymphadenopathy syndrome (LAS)

Non-nucleoside reverse transcriptase inhibitors (NNRTI)
Nucleoside reverse transcriptase inhibitors (NRTI)
Reverse transcriptase
T cell
Zidovudine (AZT)

Microorganisms to Know

Cytomegalovirus
Epstein-Barr virus (EBV)
Human herpesvirus-8 (HHV-8)
Human immunodeficiency virus (HIV) types 1 and 2
Human papillomovirus (HPV)
Mycobacterium avium complex (MAC)

Includes:
Mycobacterium avium
  Mycobacterium intracellulare
Mycobacterium tuberculosis
Pneumocystis carinii
Toxoplasma gondii

Diseases to Know

AIDS
B cell lymphoma
Cytomegalovirus disease
HIV disease
Kaposi's sarcoma

MAC disease
Mycobacterial disease
Pneumocystosis
Toxoplasmosis
Thrush

Learning Activities

1. On the drawing of the human immunodeficiency virus (HIV) label the following:

Lipid envelope, Surface glycoprotein, Transmembrane protein, Matrix protein, Capsid protein, Reverse transcriptase, Integrase, Protease, Nucleocapsid, RNA, HLA antigens

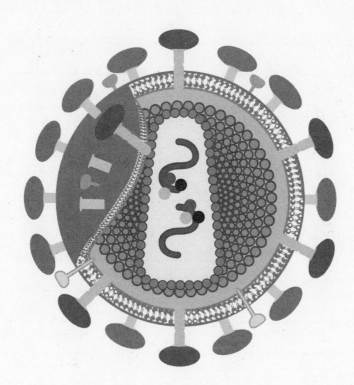

2. Explain why HIV disease is important in terms of incidence (new cases in the population) and prevalence (total cases in the population). How has the incidence and prevalence of HIV disease changed in the past 15 years?

3.   Identify the causative agent of HIV disease.

4.   Describe the seven major symptoms of HIV disease that constitute the acute retroviral syndrome. (See text, Table 29.5)

1.
2.
3.
4.
5.
6.
7.

5.   What type of cells does HIV infect? Why does it infect these cells?

6.   List three ways that HIV is transmitted.

1.
2.
3.

7.   List five major risk factors for HIV disease.

1.
2.
3.
4.
5.

8. List nine lifestyles that help to control HIV disease.

1.	
2.	
3.	
4.	
5.	
6.	
7.	
8.	
9.	

9. Why have vaccines not been successful in the prevention of HIV disease? What prospects are under investigation?

10. List four kinds of tumors that are associated with HIV disease.

1.	
2.	
3.	
4.	

11. Describe four major infectious complications of acquired immunodeficiency. Indicate cause, pathology and treatment.

Complication	Cause	Pathology	Treatment

Self Test

1. The variability in HIV virions is due to frequent errors made by reverse transcriptase in copying the viral genome.

    a. true
    b. false

2. The incubation period for HIV is

    a. 10 years.
    b. 2 to 6 months.
    c. 6 days to 6 weeks.
    d. 2-3 weeks.
    e. 2-3 days.

3. Subtypes of HIV-1, the causative agent of HIV infection and AIDS, vary in their

    a. geographical distribution.
    b. ability to infect by different routes.
    c. host species.
    d. Both a and b are correct.
    e. Both a and c are correct.

4. Symptoms of HIV infections are often

    a. asymptomatic.
    b. mild and similar to the "flu".
    c. severe, but subside within six weeks.
    d. All three of the above are correct.
    e. Both b and c are correct.

5. HIV is _____ contagious than childhood chickenpox.

    a. much more
    b. about as
    c. much less

6. The AIDS epidemic is increasing most rapidly among

    a. homosexual males.
    b. heterosexuals.
    c. intravenous drug abusers.
    d. Both a and c are correct.
    e. Both b and c are correct.

7. Most AIDS victims die from

    a. pneumocystosis.
    b. toxoplasmosis.
    c. tuberculosis.
    d. Kaposi's sarcoma.
    e. lymphoma.

8. Cytomagalovirus is a common cause of which of the following complications in AIDS?

    a. dementia
    b. tuberculosis
    c. blindness
    d. lymphoma
    e. Kaposi's sarcoma

9. For which of the following diseases should individuals with HIV disease be immunized?

    a. measles, mumps and rubella
    b. hepatitis B
    c. varicella-zoster
    d. All of the above.
    e. Only a and b.

10. The effectiveness of treatment HIV disease is limited by

    a. the high mutation rate of HIV.
    b. side effects of medications.
    c. development of resistance by HIV.
    d. All of the above are correct.
    e. Only a and b are correct.

Thought Questions

1. Why has it been so difficult to produce a vaccine to prevent HIV disease?

2. Describe what can be done to reduce the risk of HIV infection in a given population.

3. Describe the treatment and effectiveness of treatment of HIV disease. Explain why HIV disease
    is difficult to treat.

Answers to Self Test Questions

1-a, 2-c, 3-d, 4-e, 5-c, 6-e, 7-a, 8-c, 9-e, 10-d

Chapter 30   Microbial Ecology

Overview

Microorganisms are found throughout the biosphere and are essential for life as we know it. They both change the environment in which they are found and adapt to that environment. Those microorganisms that are best adapted tend to dominate in a particular environment. The roles of microorganisms are many and varied including primary production, decomposition and biogeochemical cycling.

Learning Objectives

After studying the material in this chapter, you should be able to:

1. Define the following terms:
    *   Ecology
    *   Community
    *   Ecosystem
    *   Ecological niche
    *   Biosphere
    *   Biodiversity
    *   Microenvironment
    *   Macroenvironment
    *   Biofilms
2. Describe the roles of:
    *   Primary producers
    *   Consumers
    *   Decomposers
3. Describe how microorganisms grow in low-nutrient environments.
4. Distinguish between microbial competition and antagonism.
5. Describe how environmental changes have an effect on microbial communities.
6. Describe the roles of the following techniques or procedures in studying microbial ecology
    *   Fluorescence *in situ* hybridization (FISH)
    *   Polymerase chain reaction (PCR)
    *   Denaturing gradient gel electrophoresis (DGGE)
    *   Genomics
7. Distinguish between oligotrophic and eutrophic waters.
8. Compare marine environments with freshwater environments.
9. Describe the effects of nutrient-rich runoff on water.
10. Define the following:
    *   Groundwater
    *   Salt lakes
11. List the characteristics of soil
12. Describe the kinds of microorganisms that inhabit the soil and list their roles.
13. Define biogeochemical cycles and list three specific examples.
14. List three uses of chemical elements in biogeochemical cycles.

15. Describe what occurs in each of the following cycles.
    - Carbon cycle
    - Nitrogen cycle
    - Sulfur cycle
    - Phosphorus cycle
16. List two sources of energy for ecosystems.
17. Define mutualism.
18. Describe the two major types of mycorrhizal relationships.
19. Differentiate among nitrification, denitrification and ammonification.
20. Describe nitrogen fixation and explain its importance.
21. Describe the relationship between microorganisms and herbivores.

Key Concepts

1. Microorganisms are found throughout the biosphere.
2. The best-adapted organism takes over its microenvironment.
3. Primary producers in aquatic environments convert, through the process of photosynthesis, inorganic substances such as CO_2 to organic matter using energy obtained from sunlight.
4. Microorganisms can change their environments and can adapt to environmental changes.
5. Water contains a wide variety of microorganisms, which vary greatly with the different kinds of aquatic environments.
6. The soil contains a wide variety of microorganisms that are essential for modifying, degrading and producing biologically important substances.
7. Environmental factors such as moisture, pH, temperature, and nutrient supply affect the numbers and kinds of microorganisms in the soil.
8. Energy for ecosystems can come from either sunlight via photosynthesis or from the synthesis of inorganic or organic compounds by chemotrophic microorganisms.
9. Microorganisms are essential in the recycling of biologically important elements such as oxygen, carbon, nitrogen, sulfur and phosphorus.

Summary Outline

19.1 Principles of microbial ecology
   A. Ecosystems vary in their **biodiversity** and **biomass**.
   B. The microenvironment of a microorganism is most important to its survival and growth.
   C. Nutrient acquisition
      1. **Primary producers** convert carbon dioxide into organic materials.
      2. **Consumers** use the organic materials, either directly or indirectly, produced by plants.
      3. **Decomposers** breakdown the remains of primary producers and consumers.
   D. Bacteria in **low nutrient environments** grow in dilute aqueous solutions often in **biofilms**.
   E. Microbial competition and antagonism
      1. Microorganisms in the environment compete for the same limited pool of nutrients.
      2. One species may competitively exclude others, or produce compounds that inhibit others.
   F. Environmental changes, which are common, cause changes in microorganisms that include induction of enzymes, selection of mutants, and changes in dominance.

    G. Microorganisms often grow in communities attached to a solid substrate or at air-water interfaces.

    H. Microbial ecology is difficult to study because few environmental microorganisms can be grown in the laboratory, but molecular techniques such as **fluorescence *in situ* hybridization, polymerase chain reaction (PRC), denaturing gradient gel electrophoresis (DGGE)**, and **DNA sequencing** are being used to understand complex microbial communities.

19.2 Aquatic habitats

    A. Types of environments

      1. **Oligotrophic waters** are nutrient poor.

      2. **Eutrophic waters** are nutrient rich.

      3. Overgrowth of aerobic heterotrophs can cause an aquatic environment to become **hypoxic** resulting in the death of aquatic animals.

      4. Marine environments are usually oligotrophic and aeobic, but inshore areas can be affected by nutrient runoff.

      5. Freshwater environments

        a. Oligotrophic lakes may have anaerobic layers due to thermal stratification.

        b. Shallow, turbulent streams are generally aerobic.

      6. Specialized environments – Salt lakes, mineral-rich and hot springs support the growth of microorganisms that are adapted to survive in these special environments.

19.3 Terrestrial habitats

    A. Soil constitutes and environment that can fluctuate greatly.

    B. The density and composition of soil microorganisms are affected by environmental conditions.

    C. The most important **environmental influences** in soil are: **Moisture, acidity, temperature, and nutrient availability**

    D. The **rhizosphere** is the zone of soil that adheres to plant roots and it contains a much higher concentration of microorganisms than the surrounding soil.

19.4 Biogeochemical cycling and energy flow

    A. All organisms use elements to produce **biomass**, as a **source of energy**, and as a **terminal electron acceptor**.

    B. **Carbon cycle**—The carbon cycle revolves around CO_2, its fixation into organic compounds by primary producers, and its regeneration mostly by microorganisms.

    C. **Nitrogen cycle**—Atmospheric nitrogen is converted to biologically useful forms through the processes of **ammonification, nitrification, denitrification,** and **nitrogen fixation** by free-living and symbiotic nitrogen fixers.

    D. **Sulfur cycle**—The sulfur cycle is similar to the nitrogen cycle.

    E. **Energy sources** for ecosystems include (1) **sunlight** via photosynthesis and (2) the **chemical synthesis of inorganic and organic materials** by chemoautotrophic microorganisms.

19.5 Mutualistic relationships between microorganisms and eukaryotes

    A. Mycorrhizae are fungi that help plants to take up phosphorus and other substance from the soil while gaining nutrition for their own use.

    B. Symbiotic nitrogen-fixing microorganisms add a significant amount of fixed nitrogen to the soil.

    C. A mutualistic relationship exists between microorganisms and herbivores that aids in the digestion of plant material.

Terms You Should Know

Ammonification
Antagonism
Antagonism
Bacteriocins
Biodiversity
Biofilm
Biomass
Bioremediation
Biosphere
Commensalism
Community
Competition

Consumer
Decomposer
Denitrification
Ecological niche
Ecology
Ecosystem
Eutrophic waters
Hypoxic
Macroenvironment
Microbial mat
Microbial mat
Microenvironment

Mutualism
Mycorrhizae
Nitrification
Nitrogen fixation
Oligotrophic waters
Primary producer
Putrefaction
Rhizosphere
Salt lakes
Thermal stratification

Microorganisms to Know

Streptococcus lactis
Lactobacillus casei

Lactobacillus bulgaricus
Rhizobium

Learning Activities

1. Match the following terms with their definitions:

	Ecology	A.	The role that an organism plays in its ecosystem
	Community	B.	Organisms that use CO_2 as a carbon source and convert it to organic compounds
	Ecosystem	C.	Conversion of ammonium ions to nitrate
	Ecological niche	D.	Conversion of gaseous nitrogen to nitrate or ammonium
	Biosphere	E.	Conversion of nitrate to gaseous nitrogen
	Primary producers	F.	The study of the relationship of organisms to each other and to their environment
	Nitrification	G.	Biological cleanup of pollutants in the environment
	Denitrification	H.	Living organisms in a given area
	Ammonification	I.	Constituted by living organisms interacting with their environment
	Nitrogen fixation	J.	The production of ammonia during the catabolism of protein
	Bioremediation	K.	Process by which light energy is converted to chemical bond energy
	Photosynthesis	L.	Conversion of organic nitrogen into ammonium ions

2. How do microorganisms deal with a low-nutrient environment?

3. How do competition and antagonism affect a community?

4. Describe the effects of environmental change on microbial communities.

5. List four techniques or procedures used to study microbial ecology.

6. Distinguish between:

Oligotrophic and eutrophic waters	
Marine and freshwater environments	

7. List four environmental influences that affect the density and composition of soil flora.

1.	
2.	
3.	
4.	

8. List two ways that ecosystems can obtain energy.

1.
2.

9. List three ecological roles of microorganisms in recycling.

1.
2.
3.

10. Briefly describe what occurs in each of the following cycles.

Carbon cycle	
Nitrogen cycle	
Sulfur cycle	

11. Explain why nitrogen fixation is important.

Self Test

1. All of the living organisms in a given area form

   a. a community.
   b. an ecosystem.
   c. an ecological niche.
   d. a biosphere.
   e. a biomass.

2. Primary producers include:

   1. Plants
   2. Algae
   3. Cyanobacteria
   4. Protozoa

   a. 1, 2 and 3 are correct.
   b. 2 and 4 are correct.
   c. 1 and 3 are correct.
   d. 1, 2, 3 and 4 are correct.
   e. All are correct.

3. An example of microbial antagonism is the production of

   a. bacteriocins.
   b. mycorrhizae.
   c. geosmins.
   d. nod factors.

4. Primary producers

   a. use CO_2 as a carbon source.
   b. convert CO_2 into an organic compound.
   c. are the first step in the food chain.
   d. All of the above are correct.
   e. Only a and c are correct.

5. Aquatic habitats that are nutrient poor are termed

   a. eutrophic.
   b. oligotrophic.
   c. marine.
   d. fresh.

6. Soil microorganisms include which of the following?

   1. Bacteria
   2. Fungi
   3. Algae
   4. Protozoa
   5. Multicellular parasites

   a. 1, 2 and 3 are correct.
   b. 2 and 4 are correct.
   c. 1 and 3 are correct.
   d. 1, 2, 3 and 4 are correct.
   e. All are correct.

7. All of the following are environmental influences that affect soil except

   a. moisture.
   b. acidity.
   c. light.
   d. temperature.
   e. nutrient availability.

8. The largest group of bacteria that inhabit the soil are

   a. actinomycetes.
   b. *Clostridium*.
   c. *Azotobacter*.
   d. *Bacillus*.
   e. *Agrobacterium*.

9. In biogeochemical processes elements are used

   a. for biomass production.
   b. as an energy source.
   c. as terminal electron acceptors.
   d. All of the above are correct.

10. The process by which nitrogen is made available for use by plants is

    a. ammonification.
    b. nitrification.
    c. denitrification.
    d. nitrogen fixation.
    e. None of the above.

Thought Questions

1. Why might herbicides and pesticides be harmful to the ecology of the soil?

2. If microorganisms are capable of fixing nitrogen, why is it necessary for farmers to use fertilizers?

Answers to Self Test Questions

1-a, 2-a, 3-a, 4-d 5-b, 6-a, 7-c, 8-a, 9-d, 10-d

Chapter 31   Environmental Microbiology: Treatment of Water, Wastes and Polluted Habitats

Overview

A healthy environment requires clean water. Water supplies are continually treated and tested to avoid contamination. Modern society produces waste materials at an ever-increasing rate. As waste materials buildup in the environment of any organism, its health is negatively affected. Waste water and sewage treatment methods are covered in this chapter, as well as methods for treatment and disposal of solid wastes. The concept of bioremediation is introduced and discussed.

Learning Objectives

After studying the material in this chapter, you should be able to:

1. Define biochemical oxygen demand (BOD) and explain its significance in sewage treatment.
2. Describe the goals of sewage treatment.
3. Describe the physical, chemical and biological processes used in sewage treatment.
4. List uses for treated waste residues.
5. Describe or identify the following sewage treatment methods:
    *   lagooning
    *   trickling filters
    *   septic tanks
    *   artificial wetlands
6. Explain how water can be safe to drink while still containing microorganisms.
7. List three steps in the treatment of water in order to make it safe for drinking.
8. Explain what is indicated by the presence of coliforms in water.
9. List four procedures used to test for coliforms in water.
10. List the advantages and disadvantages of using landfills for solid waste disposal.
11. Define bioremediation and describe how it can be accomplished.

Key Concepts

1. There is a direct relationship between the health of a population and its environment.
2. Proper sewage treatment, necessary to ensure the health of a community, depends on the stabilization of wastes by microorganisms.
3. Primary sewage treatment is physical and involves the removal of large objects and some of the particulate matter through screening and sedimentation.
4. Secondary sewage treatment is biological and involves the conversion of organic materials to inorganic and reduction of the BOD of the sewage.

5. Tertiary sewage treatment is chemical and biological, and involves the removal of nitrates and phosphates. Phosphates are precipitated and nitrates are biologically removed by bacteria such as species of *Pseudomonas* and *Bacillus*.
6. Pathogenic bacteria are usually eliminated by secondary sewage treatment.
7. Methods such as septic tanks, trickling filters and the production of artificial wetlands are good solutions to small-scale sewage treatment problems.
8. Adequate water treatment and regular testing help to assure safe drinking water and recreational waters.
9. Both commercial and backyard composting reduce the need for large landfills for disposal of solid waste.
10. Bioremediation is the use of biological agents such as bacteria and fungi to degrade or detoxify pollutants.

Summary Outline

13.1 Microbiology of sewage treatment
    A.   **Reduction of biochemical oxygen demand (BOD)** reflects the effectiveness of treatment.
    B.   Multiple sewage treatment methods
        1. **Primary treatment** of sewage is a physical process designed to remove materials that sediment out.
        2. **Secondary treatment** is chiefly a process designed to convert most of the suspended solids to inorganic compounds and microbial biomass removing most of the BOD
        3. **Tertiary treatment** is generally designed to remove nitrates and phosphates.
        4. **Biosolids** that result from anaerobic digestion of **sludge** can be used to improve soils and promote plant growth.
    C.   Individual sewage treatment systems -Rural dwelling rely on septic tanks for sewage disposal
13.2 Drinking water treatment and testing
    A.   Water treatment processes
        1. Metropolitan water supplies are treated to remove **particulate** and **suspended matter**, various **microorganisms**, and **organic waste**.
        2. Water is treated using chlorine or other disinfectants to kill harmful bacteria and viruses.
    B.   Water testing – **Coliforms** are used as **indicator organisms**, suggesting the possible presence of pathogens.
13.3 Microbiology of solid waste treatment
    A. Sanitary **landfills** for solid waste disposal – landfills are used to dispose of solid wastes near towns and cities; disadvantages include limited available sites and slow decomposition of wastes
    B. Commercial and backyard **composting** – an alternative to landfills – composting offers cities a way to reduce the amount of garbage sent to landfills.
13.4 Microbiology of **bioremediation**
    A. Pollutants
        1. **Pollutants** that are concentrated in a new environment can remain for years.
        2. Synthetic compound are more likely to be **biodegradable** if they have a chemical composition similar to that of naturally occurring compounds.
    B. Means of bioremediation include using nitrogen and phosphorus containing fertilizers to increase the effectiveness of oil degradation by naturally occurring bacteria.

Terms You Should Know

Activated sludge process
Aquifer
Artificial wetlands
Biochemical oxygen demand
    (BOD)
Bioremediation
Coliform bacteria
Effluent

Ground water
Indicator organism
Lagoons
Most probable number method
Pollutants
Potable
Receiving water
Sanitary landfills

Septic tank
Sewage
Sludge
Surface water
Total coliform rule
Trickling filter
Watershed
Xenobiotics

Microorganisms to Know

Clostridium species
Cryptosporidium species
Enterococci

Escherichia coli
Escherichia coli O157:H7
Giardia lamblia

Thiothrix species

Learning Activities

1. Explain why there is a direct relationship between the environment and the health of the human population.

2. Define biochemical oxygen demand (BOD) and explain its significance in sewage treatment.

3. Explain why and how phosphates and nitrates are removed during tertiary sewage treatment.

4. Match the method with its function.

	Primary sewage treatment	a.	Sewage is channeled into shallow ponds, where settling occurs and sewage materials are stabilized by anaerobic and/or aerobic organisms
	Secondary sewage treatment	b.	Conversion of organic materials to inorganic and reduction of the BOD of sewage
	Tertiary sewage treatment	c.	Sewage wastes are channeled into a revolving arm, and through holes in the bottom of the arm onto a gravel and rock bed, which is coated with microorganisms that convert organic material into inorganic material
	Lagooning	d.	Sewage is channeled into successive ponds in which aerobic and anaerobic stabilization occurs
	Trickling filters	e.	Sewage is collected in a tank in which much of the solid material settles and is degraded by anaerobic microorganisms
	Septic tanks	f.	Removal of material by sedimentation
	Artificial wetlands	g.	Removal of nitrates and phosphates

5. List and describe three steps in the water treatment process.

6. Define coliforms and explain their significance in water.

7. List four methods for testing for the presence of coliforms in water.

8. List and describe two methods of solid waste treatment.

Method	Description

9. Describe the function of bioremediation.

10. List and describe two general strategies for bioremediation.

Self Test

1. Every day, each American produces an average of _____ gallons of sewage.

   a. 1.2
   b. 12
   c. 120
   d. 1200
   e. 5

2. The biological oxygen demand (BOD) roughly measures the amount of _____ in water.

   a. degradable organic material
   b. oxygen
   c. degradable inorganic material
   d. All of the above
   e. Only a and c are correct.

3. Place the following water treatment procedures in the proper order.

   1. Water is treated with chlorine or other disinfectants.
   2. Water is allowed to stand in order for particulate matter to settle.
   3. Water is filtered through sand or diatomaceous earth.
   4. Water is mixed with a flocculent chemical.
   5. Flocculation occurs.

   a. 1, 2, 3, 4, 5
   b. 2, 3, 4, 5, 1
   c. 4, 5, 3, 2, 1
   d. 2, 4, 5, 3, 1
   e. 2, 4, 5, 1, 3

4. The activated sludge process is _____ sewage treatment method.

   a. a primary
   b. a secondary
   c. a tertiary
   d. not a

5. Which of the following treatments removes phosphates and nitrates from sewage?

   a. primary
   b. artificial wetlands
   c. secondary
   d. digester
   e. tertiary

6. During which of the following treatments are pathogens generally eliminated?

   a. primary
   b. artificial wetlands
   c. secondary
   d. digester
   e. tertiary

7. Which of the following is not a good solution for individual sewage treatment problems?

   a. septic tanks
   b. trickling filters
   c. lagooning
   d. artificial wetlands
   e. All of the above are good solutions.

8. Coliforms

   a. are Gram-negative bacteria.
   b. include *Escherichia coli*.
   c. ferment lactose.
   d. are often found in soil and in the gut and feces of warm-blooded animals.
   e. All of the above are true.

9. Which of the following are useful solutions for solid waste disposal?

   a. sanitary landfills
   b. ocean dumping
   c. composting
   d. Both a and c
   e. All of the above.

10. The use of biological agents to degrade or detoxify pollutants is known as

   a. Xenobiotics
   b. Bioaugmentation
   c. Bioventing
   d. Bioremediation
   e. Biostimulation

Thought Questions

1. What is biochemical oxygen demand (BOD) and why is it important in sewage treatment?

2. Describe the role of each of the three steps in large-scale treatment of sewage.

Answers to Self Test Questions

1-c, 2-a, 3-d, 4-b, 5-e, 6-c, 7-a, 8-e, 9-d, 10-d

Chapter 32   Food Microbiology

Overview

Microorganisms play important roles in both the production and spoilage of foods and beverages. Through the process of lactic acid fermentation by bacteria, cheese, yogurt, acidophilus milk, pickled vegetables and fermented meat products such as salami, pepperoni and summer sausage are produced. Alcoholic fermentation by yeasts is used to produce wine, beer, distilled spirits, vinegar and bread. Mold growth is used to produce soy sauce and to give flavor and texture to some cheeses. Microorganisms cause destruction of food both as crops and as stored items. Our knowledge of the processes of spoilage can enable us to inhibit these processes and preserve foods. Microorganisms are responsible for food-borne illness. Food-borne intoxication results from the consumption of toxin produced by a microorganism growing on food as seen in botulism. Food-borne infections require the consumption of living microorganisms. Examples of causative agents include *Salmonella, Campylobacter* and *Escherichia coli* O157:H7. This chapter presents the principles of food microbiology.

Learning Objectives

After studying the material in this chapter, you should be able to:

1. List examples of foods or beverages that are produced by microorganisms.
2. Distinguish between fermentation and spoilage.
3. Explain why the fermentation process often prevents foods or beverages from spoiling.
4. List intrinsic and extrinsic factors that influence the growth of microorganisms in foods.
5. List three major ways in which microorganisms produce food and give examples of each.
6. List seven common food spoilage bacteria.
7. List four common food spoilage fungi.
8. List ten common causes of food-borne infections.
9. List two common causes of food-borne intoxications.
10. Describe how food-borne infections and intoxications can be prevented.
11. List nine methods of food preservation and give examples of their uses.

Key Concepts

1. Many different kinds of microorganisms can use food as a growth medium. The end products that they produce can be desirable (fermented foods), undesirable (spoilage) or harmful (food-borne disease).
2. The types of microorganisms that grow and predominate in a food product can be influenced by intrinsic factors such as moisture, pH, and the presence of antimicrobial chemicals, and by extrinsic factors such as temperature and atmosphere.
3. The metabolic activities of bacteria and yeasts can be used to produce a wide variety of foods and beverages including cheese, yogurt, sauerkraut, bread, wine and beer. The three major processes are (1) lactic acid fermentation by the lactic acid bacteria, (2) alcoholic fermentation by yeast and (3) changes due to mold growth.

4. Metabolic products of microorganisms can spoil foods by producing undesirable flavors, odors and textures.
5. Food-borne intoxication results from the consumption of toxins produced by microorganisms growing on a food.
6. Food-borne infection results from consumption of living microorganisms.
7. Food spoilage can be prevented or retarded by destroying microorganisms or by altering their environmental conditions to inhibit growth.

Summary Outline

32.1 **Principles of food microbiology**
   A. Food is an **ecosystem** in which microorganisms compete to metabolize the nutrients, making **end products** such as **acids, alcohols and gas**.
   B. Foods that have been altered by controlling the activity of bacteria, yeasts or molds are called **fermented**.
   C. Biochemical changes in foods, when perceived as unpleasant are called **spoilage**.
   D. Growth of pathogens generally does not result in perceptible changes in quality of food, but can result in **food-borne disease**.
   E. **Spoilage** can be delayed or prevented and **food-borne illness** can be avoided by slowing the growth of microorganisms, or by reducing or eliminating the initial numbers of them on foods

32.2 **Factors influencing the growth of microorganisms in food**
   A. **Intrinsic** and **extrinsic factors** determine which microorganisms can grow and predominate in a food product.
   B. **Intrinsic factors**
      1. **Bacteria require a high a_w (water activity)**. They grow quickly on fresh, moisture-rich foods but not on dry, sugary or salted foods. **Fungi** can grow on foods which have an a_w **too low to support the growth of bacteria**.
      2. The **pH of a food** is important in determining which organisms can survive and thrive on it; many species of **bacteria**, including most pathogens, are **inhibited by acidic conditions**.
      3. **Coverings** help **protect** some foods from the **invasion of microorganisms**.
      4. Some foods contain **natural antimicrobial chemicals** that may help prevent spoilage.
   C. **Extrinsic factors**
      1. **Low temperatures halt or inhibit the growth** of most food-borne microorganisms.
      2. The **presence or absence of oxygen** impacts the type of microbial population able to grow on a food.

32.3 **Microorganisms in food and beverage production**
   A. The **acids produced in fermented foods inhibit** the growth of many **spoilage organisms** as well as **food-borne pathogens**.
   B. The tart taste of yogurt, pickles, sharp cheese and some sausages is due to the production of **lactic acid** by species of *Lactobacillus, Lactocossus, Streptococcus, Leuconostoc* and/or *Pediococcus*.
   C. **Alcoholic fermentations**: The yeast *Saccharomyces* **ferments sugar to produce ethanol and carbon dioxide** in the production of **wine, beer**, and **distilled spirits**. **Vinegar** is the product of the **oxidation of alcohol** by *Acetobacter* and *Gluconobacter*. In bread-making, the **CO_2 produced by yeast causes bread to rise**, and the **alcohol is lost to evaporation**.
   D. Changes imparted by mold: Some **cheeses and other dishes** are produced by **encouraging the growth of molds** on foods. **Soy sauce** is made by allowing species of *Aspergillus* to degrade a mixture of **soybeans and wheat**, which is then **fermented in brine**.

32.4 **Food spoilage** is often due to the metabolic activities of microorganisms as they grow and utilize the nutrients in the food.

   A.   Bacteria: **Psychrophilic species of *Pseudomonas*** can multiply at refrigeration temperatures and metabolize a wide variety of compounds, causing spoilage of foods including meats and vegetables. Other important causes of food spoilage include ***Ervinia, Acetobacter, Alcaligenes,*** **lactic acid bacteria** and **endospore formers**.

   B.   **Fungi** grow in **acidic** and **low moisture environments**, therefore fruits and breads are more likely to be spoiled by fungi than by bacteria.

32.5 **Food-borne illness**

   A.   **Food-borne intoxication** is an illness that results from the consumption of a **toxin** produced by a microorganism growing in a food product. Strains of ***Staphylococcus aureus*** produce a toxin that can cause nausea and vomiting. The anaerobic, spore-forming, Gram-positive rod ***Clostridium botulinum*** produces a **neurotoxin**, which can be destroyed by boiling for 10-15 minutes.

   B.   **Food-borne infection** requires the **consumption of living organisms**. Cooking of food immediately before consumption prevents food-borne infection. ***Salmonella, Campylobacter*** **species** and ***E. coli* O157:H7** are significant causes.

32.6 **Food preservation** can be accomplished by **killing microorganisms** or **altering conditions to inhibit their growth.** Methods used to preserve foods include **canning, pasteurization, cooking, freezing, refrigeration, reducing the a$_w$, lowering the pH, adding antimicrobial chemicals,** and **irradiation**.

Terms You Should Know

Aflatoxin	Food-borne intoxication	Rennin
Baker's yeast	Hops	Spoilage
Botulism	Intrinsic factors	Starter cultures
Curd	Mashing	Water activity (a$_w$)
Extrinsic factors	Pasteurization	Whey
Food-borne illness	Pickling	Wort
Food-borne infection	Preservatives	

Microorganisms to Know

Involved in food production

	Lactococcus lactis	*Saccharomyces cerevisiae*
	Leuconostoc mesenteroides	*Aspergillus oryzae*
Lactobacillus brevis	*Pediococcus species*	*Aspergillus sojae*
Lactobacillus plantarum	*Penicillium camemberti*	*Acetobacter species*
Lactobacillus acidophilus	*Penicillium candidum*	*Gluconobacter species*
Lactobacillus bulgaricus	*Penicillium roquefortii*	*Streptococcus thermophilus*
Lactococcus cremoris	*Propionibacterium shermanii*	

Involved in food spoilage

Acetobacter species	*Bacillus stearothermophilus*	*Leuconostoc species*
Alcaligenes species	*Botrytis species*	*Streptococcus species*
Alernaria species	*Erwinia species*	*Penicillium species*
Aspergillus flavus	Lactic acid bacteria	*Pseudomonas species*
Bacillus coagulans	*Lactobacillus species*	*Rhizopus species*

Involved in food-borne illness

Campylobacter species

Clostridium botulinum
Clostridium perfringens
Escherichia coli 0157:H7
Listeria monocytogenes

Salmonella species
Shigella species
Staphylococcus aureus
Vibrio parahaemolyticus

Learning Activities

1. Distinguish between fermentation and spoilage.

2. Why does the fermentation process prevent some foods or beverages from spoiling?

3. List six intrinsic factors that influence the growth of microorganisms.

1.	
2.	
3.	
4.	
5.	
6.	

4. List two extrinsic factors that influence the growth of microorganisms in foods.

1.	
2.	

5.   List five foods produced using lactic acid bacteria and name the specific microorganism involved.

Foods	Microorganism

6.   List the causative microorganism and describe the role of alcohol fermentation in

Food/Beverage	Microorganism and its role
Wine	
Sake	
Beer	
Distilled spirits	
Vinegar	
Breads	

7.   List four foods produced using molds and name the specific microorganism involved.

Foods	Specific microorganism

8. List seven common food spoilage bacteria.

1.	
2.	
3.	
4.	
5.	
6.	
7.	

9. List four common food spoilage fungi.

1.	
2.	
3.	
4.	

10. Distinguish between food-borne infections and food-borne intoxications.

11. The following is a list of common causes of food-borne intoxication and infection. Describe the symptoms and indicate a likely source for each.

Food-borne infection	Symptoms	Source
Intoxication		
Clostridium botulinum		
Staphylococcus aureus		
Infection		
Bacillus cereus		
Campylobacter species		
Clostridium perfringens		
Escherichia coli O157:H7		
Listeria monocytogenes		
Salmonella species		
Shigella species		
Vibrio parahaemolyticus		
Vibrio vulnificus		
Yersinia entercolitica		

12. How can food-borne infections and intoxications be prevented?

13. List nine methods of food preservation and give examples of their uses.

Methods	Uses

Self Test

1. Which of the following is/are essential for the growth of all microorganisms in food and beverages?

    a. vitamins
    b. oxygen
    c. water
    d. sugars
    e. darkness

2. Bacteria with an optimum growth temperature around 10°C are capable of causing spoilage in a refrigerator, and are called

    a. thermophiles.
    b. mesophiles.
    c. acidophiles.
    d. halophiles.
    e. psychrophiles.

3. Which group of bacteria is most likely to spoil a freshwater trout caught in a cold mountain stream and preserved in salt?

    a. psychrophiles
    b. halophiles
    c. anaerobes
    d. thermophiles
    e. mesophiles

4. Salts and sugars preserve foods because they

    a. make them acidic.
    b. produce a hypotonic environment.
    c. deplete nutrients.
    d. produce a hypertonic environment.
    e. provide nutrients.

5. Acidic foods are more likely to be spoiled by

   a. bacteria.
   b. fungi.
   c. acid-sensitive organisms.
   d. thermophiles.
   e. Acidic foods are rarely spoiled.

6. Which of the following bacteria are important in the production of cheese?

   a. *Lactococcus cremoris*
   b. *Lactococcus lactis*
   c. *Lactococcus bulgaricus*
   d. *Lactococcus acidophilus*
   e. Both a and b are correct.

7. Which of the following are produced through alcohol fermentation by *Saccharomyces*?

   a. wine
   b. beer
   c. distilled spirits
   d. bread
   e. All of the above.

8. *Penicillium* is used in the production of

   a. soy sauce.
   b. vinegar.
   c. miso.
   d. wine.
   e. Roquefort cheese.

9. Which of the following bacteria cause food spoilage?

   a. *Pseudomonas* species
   b. *Ervinia* species
   c. *Acetobacter* species
   d. *Streptococcus* species
   e. All of the above.

10. Which of the following causes food-borne intoxication?

   a. *Clostridium botulinum*
   b. *Escherichia coli* O157:H7
   c. *Vibrio parahameolyticus*
   d. *Yersinia enterocolitica*
   e. None of these; they all cause food-borne infections.

Thought Questions

1. *Saccharomyces* causes both the fermentation of grape juice to produce wine and the rising of bread. Why is there no alcohol in bread?

2. Freezing stops all growth of microorganisms and prevents spoilage. Why are foods damaged if they remain frozen for long periods of time?

Answers to Self Test Questions

1-c, 2-e, 3-b, 4-d, 5-b, 6-e, 7-e, 8-e, 9-e, 10-a

Notes

Notes

Notes

Notes

Notes

Notes

Notes

Notes

Notes

Notes

Notes

Notes

Notes